Experimental Evaluation of GFRP Composite Pipes

Blesma

TABLE OF CONTENTS

CHAPTER NO.	TITLE	PAGE NO.

CHAPTER NO.	TITLE	PAGE NO.

CHAPTER NO.	TITLE	PAGE NO.

CHAPTER 1

INTRODUCTION

PREAMBLE

[illegible]

COMPOSITE MATERIALS

[illegible]

CHAPTER 1

INTRODUCTION

1.1 PREAMBLE

Based on the needs, Man has created a variety of novel materials based on his end users' unique needs. The use of materials varies depending on the location. We anticipate that the material will exhibit enhanced performance qualities. The materials created must be appropriate for high-end uses, such as aerospace engineering, and satisfy requirements for increased dependability, efficiency, and performance. Extreme temperatures, cold temperatures, acidic conditions, vacuum conditions, strong magnetic fields, radiation environments, etc. are other scenarios in which the materials must function. Conventional materials cannot meet the needs of these situations. In recent decades, novel composite materials have arisen to fulfill the performance demands of industrial applications.

1.2 COMPOSITE MATERIALS

Composite materials typically consist of two phases: the matrix, which is a continuous bulk phase, and the reinforcement, which is a non-continuous, scattered phase. Some examples of composite materials include wood, epoxy reinforced with various fibers, thermoplastic reinforced with fillers, and numerous other materials. Engineers characterize composites as blends of dissimilar chemicals that are insoluble in each other and exist in discrete stages, either as continuous or discontinuous phases.

Because of their exceptional strength, stiffness, and resistance to corrosion, filament-wound Fiber Reinforced Polymer (FRP) nanocomposites find extensive application in the sectors of aerospace, automobile industries, construction, and marine-based applications. According to Bobba *et al.* (2017), Fiber Reinforced Polymer Composites (FRPC) tubes are also essential for the oil and gas industries. In recent years, the use of synthetic FRPC for helicopter structural components has seen a significant increase. Specifically, the weight proportion of FRPC on aircraft like the Boeing 787 and Airbus A350 is greater than 50% and 52%, respectively.

1.2.1 Classification of Composite Materials

Composite materials are generally classified as shown in Figure 1.1.

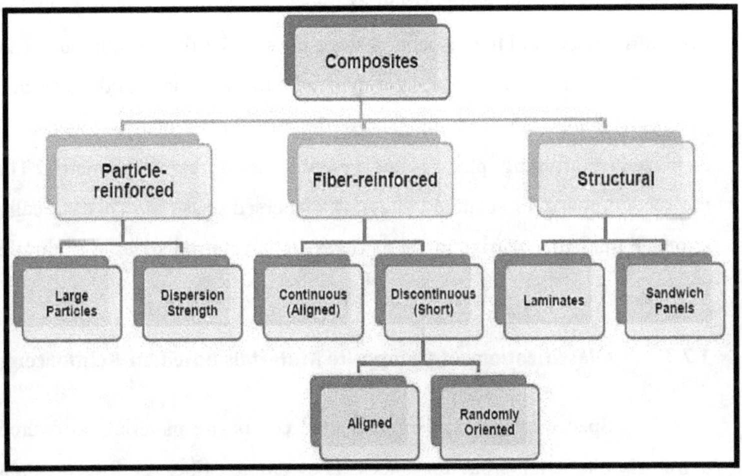

Figure 1.1 Classification of Composite Materials

"Composite" materials are commonly defined as materials that consist of two or more distinct chemical and physical phases, distinguishable by a visible interface. The several systems are painstakingly put together to create a whole with more advantageous structural or functional characteristics than any one of the parts could have by itself. Composite materials are also termed "wonder materials," and among their many advantages are their being less weight, slow to corrosion, and being easily assembled. Electronic packages, medical equipment, and aeroplanes widely use composite materials in their construction. Even though composites consist of multiple materials, each component maintains its own unique properties. Together, these distinct components provide the necessary strength or stiffness to the composite material.

Composite materials exhibit significant differences from bulk materials due to their composition of two or more distinct phases, including the matrix phase and the dispersion stage phase. The first continuous phase is called the matrix phase. Matrix is typically the less rigid and more ductile phase. It shares the load and keeps the distributed phases together. The dispersed (reinforcing) phase is not present in its entirety in the matrix. This is the second phase, or scattered phase. A dispersed phase may occasionally be stronger than the matrix; in these cases, it is referred to as a "reinforcing phase."

1.2.2 Classifications of Composite Materials based on Reinforcement

Spacecraft employ sophisticated composite materials to overcome extremely challenging circumstances. They employ fiber reinforcements with high performance within a polymer matrix, such as epoxy. Composites made up of boron and graphite are some examples. The aerospace sector limits the application of advanced composite materials. Various industries are currently using these materials. In our everyday lives, composites made of epoxy, vinyl ester, polypropylene, polyurethane, and polyester are the most commonly used. These composite materials also find applications in automobiles, turbine blades, pipes, ladder rails, the sports sector, and protective equipment.

Advantages of composite materials over traditional materials include increased flexibility, resistance to dent, ease of maintenance, and prolonged life. In many different applications, polymer composites have currently taken the place of various predictable metals and minerals. According to Kuo *et al.* (2005), composites also have advantages, including low cost, high strength, increased toughness, and light weight. Furthermore, the current work focused on a high-performance filler to create composites with enhanced properties. Figure 1.2 shows the classification of composite materials based on reinforcement.

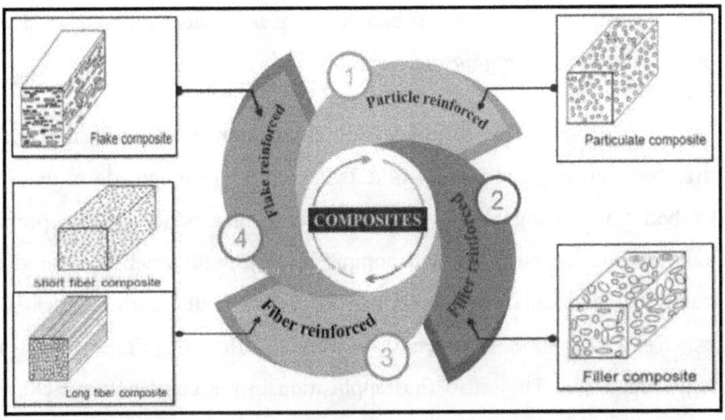

Figure 1.2 Classification based on reinforcement

1.2.3 Fiber Reinforced Composites

The matrix is important in fiber-reinforced composite material. The second stage of the composite material's development is when the fibers that merge with the matrix start to take centre stage. Because of their excellent strength, stiffness, and tribological qualities, fiber-reinforced composites find application in a variety of industrial applications.

1.2.4 Continuous Fiber Composites

Materials containing continuously dissolved fiber in the polymer matrix are known as continuous fiber-reinforced composites (CFRC). Fibers that can be utilized to reinforce polymer matrix composites include glass, aramid, and carbon fibers. Thermosetting techniques can be used to prepare these reinforced composites, which can be produced in a range of shapes and sizes Mitschang *et al.* (2003). Typically, glass fibers enhance the mechanical and thermal properties of polymer matrices such as vinyl ester resins, polyesters, epoxides, polyamides, and polypropylenes Thomason *et al.* (1997). Several objects such as boat hulls, hockey bats, tent poles, and vehicle bodywork use these composite materials.

Aramid fiber materials find application in the aeronautical sector, tires for bicycles, and mainly in defense sectors (Puttegowda *et al.* (2018)). Carbon fiber, being a light weight material, has better electric properties, enhanced tensile strength, and compressive strength when compared to the traditional materials Pimenta *et al.* (2011). Although carbon fiber is more expensive, its superior strength-to-weight ratio and rigidity lead to its widespread use. They also find application in the construction field, racing cars, and many other applications.

1.2.5 Short Fiber Composites

Fibers arranged in a discontinuous pattern make up composites reinforced having short fibres, which have a shorter length. Generally, short fibers are considered to be less than 12.5 mm in length. They are inexpensive to produce and easy to handle, which makes them suitable for production on a large scale. The advantages of short fiber composites, including their lower cost, ease of manufacturing, and isotropic nature, Ospald, *et al.* (2017) make them ideal for large-scale production. In recent years, short-fiber reinforced composites have become popular in a number of production industries. There are two types of short fiber reinforced materials: (i) randomly oriented and (ii) preferred manufacturing in the required direction.

1.2.6 Particulate Reinforced Polymer Composites

Particulate reinforced polymer composites find applications in several fields due to their ease of manufacture. Dispersion and particulate-reinforced composites are the two categories under which they fall. The particle size is small, ranging from 0.01-0.1 mm. The process of strengthening is similar to how metals harden through precipitation: the matrix shares the load, and the reinforced particulate prevents atoms from moving out of place. Usually, this strengthening occurs at the molecular level (Durmus *et al.* (2008)). These particulate-reinforced composites have applications in the sports sector, automobile industries, home appliances, electrical goods, and several fields of aviation.

1.2.7 Flake Reinforced Polymer Composites

Flat or thin flakes are attached to the polymer matrix and are frequently strengthened in a two-dimensional way within the matrix. A composite material is built up of ribbons and flakes with reinforcements arranged adjacent to each other and strengthened in both directions. In regard to efficiency, this composite material is superior to fiber-reinforced polymer composites. Conversely, composites with planar reinforcement are a relatively new idea.

1.2.8 Filler Reinforced Polymer Composites

It is possible to modify and improve the physical and mechanical characteristics of polymers by including filler reinforcements in their matrices. Fillers are a cost-effective solution that reduces the amount of matrix resin needed. Fillers enhance various properties such as conductivity, thermal and electrical resistivity, resistance to wear, friction, dielectric constant and loss, water resistance, resistance to chemical substances, microbial corrosion resistance, flame resistance, and other characteristics. The fillers can be categorized into two groups: conductive fillers and extender fillers. Utilizing conductive fillers increases both electrical and thermal conductivity, while also reducing the cost of raw materials through the use of prolonged fillers. The most common conductive and extended fillers are carbon nanotubes and different types of clay. Various sectors, including aviation, automobile appliances, electronics devices, building and construction, and the marine-based industry, use fiber-reinforced polymer composites.

1.3 CLASSIFICATION OF POLYMER COMPOSITES

Three types of polymer composites exist: macro composites, micro composites, and nanocomposites. Polymer macro composites, which are heterogeneous composites made from polymers and macro sized fillers, find extensive application in various fields. In micro and macro composites, the reinforcement of phases is millimetres long. Composites with micron-sized fibers will have fairly high aspect ratios. Micro composite materials include hollow spherical glass-filled composites and micron-sized fillers like silicon carbide and alumina, to name a few. The addition of micron-sized fillers improves the durability, rigidity, and heat characteristics of a material. The presence of macroscopic elements may cause faults in the composites created as a result of this procedure. The composite's flaw diminishes as the size of its reinforcing components decreases. "Polymer nanocomposites" are those with at least one material in the 1 nm range Kumar *et al.* (2009).

1.3.1 Thermoplastics

Thermoplastic polymers have an elongated structure with no chemical links between molecules. Heat and pressure can temporarily break the weak connections between molecules, allowing the polymers to soften and flow. When the molecules cool, they reestablish their secondary bonds, and the plastics transition from fluid to solid form. Composite products are increasingly using thermoplastics due to their affordability and ease of manufacture. The most common thermoplastic resins are polyethylene (PE), polypropylene (PP), polystyrene (PS), polyvinylchloride (PVC), polyamide (PA), polycarbonate (PC), and polymethyl methacrylate (PMMA).

1.3.2 Thermoset

A thermoset polymer consists of chemically bonded molecules that form a 3D-type structure. Once the crosslinked network forms, heat and pressure cannot melt or reflow the thermoset polymer. In general, thermoset resins exhibit greater thermal stability and heat resistance compared to other thermosets. This property enables its use in high temperature applications. The thermoset resins include polyurethanes and phenols. But there are also unsaturated polyesters, vinyl esters, and obviously epoxy. The epoxy resins must be cured at either ambient or higher temperatures.

1.4 EPOXY RESIN

In structural composites, epoxy frequently serves as a polymer matrix. This is attributed to epoxy's strong adhesive qualities as well as its extensive track record in composite materials. Epoxy offers a robust amalgamation of mechanical properties, exhibits resistance to corrosion, maintains dimensional stability, demonstrates exceptional adhesion, and is cost-effective. Moreover, the liquid form of uncured epoxide resin exhibits remarkably high molecular mobility during processing due to its low molecular weight. This fiber motion facilitates the rapid movement of the resin along the fiber's surface. Epoxy resins consist of a minimum of two epoxide groups per molecule. Figure 1.3 depicts the chemical arrangement of an epoxide group. When an epoxy, a thermosetting polymer, mixes with a catalyst, also known as a hardener, it undergoes a chemical reaction known as curing. The curing process includes polymerisation and cross-linking chemical reactions. Temperature manipulation and the selection of specific resin and hardener ingredients can regulate the polymerization procedure, also known as curing.

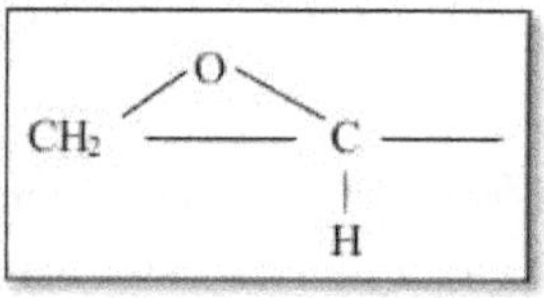

Figure 1.3 Structure of Epoxy Resin

1.4.1 Epoxy Resin Types

Epoxy resins can be divided into two types: glycidyl and non-glycidyl. Glycidyl epoxies are further classified as glycidyl-ether, glycidyl-ester, and glycidylamine. Non-glycidyl epoxy resins are either aliphatic or cycloaliphatic epoxy resins. Epichlorohydrin reacts with a suitable dihydroxy molecule or dibasic acid to form glycidyl epoxies, while an olefinic double bond peroxidizes to form nonglycidyl epoxies. Glycidyl-ether epoxies, such as diglycidyl ether of bisphenol A (DGEBA) and novolac epoxy resins, are the most commonly used.

1.4.2 Diglycidyl Ether of Bisphenol-A (DGEBA)

Bisphenol-A and epichlorohydrin combine to produce DGEBA, a commercially available epoxy resin, without the need for a catalyst. An integer of n, the number of repeating units, also known as the polymerization degree, determines the properties of DGEBA resins. The stoichiometry of the synthesis reaction determines the number of unit repetitions. In numerous commercial products, the numerical value of n frequently falls between 0 and 25. Figure 1.4 illustrates the structure of DGEBA.

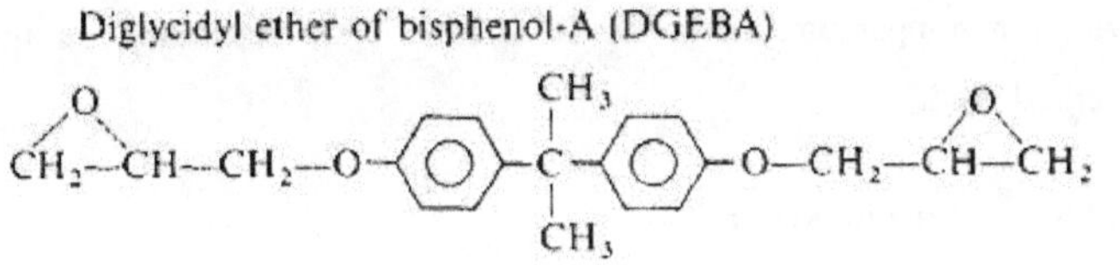

Figure 1.4 Structure of DGEBA

1.4.3 Novolac Epoxy Resins

The synthesis of Novolac epoxy resins, glycidyl ethers, begins with phenolic novolac resin. An acidic catalyst aids in the reaction of phenols with excess formaldehyde to produce phenolic novolac. A chemical reaction takes place between phenolic novolac resin and epichlorohydrin, with sodium hydroxide acting as a catalyst. This makes novolac epoxy resins. Novolac epoxy resins usually have several epoxide groups per molecule, which is dictated by the number of phenolic hydroxyl groups in the novolac resin, their reactivity, and the degree of polymerization of small molecules during production. These resins have a high number of epoxide groups, which enables them to form a dense network of chemical bonds. As a result, they exhibit exceptional resistance to heat, chemicals, and solvents. Novolac epoxy resins are extensively used in microelectronics packaging because of their exceptional performance in high-temperature conditions, outstanding moldability, favourable electrical characteristics, durability to heat and moisture, and impressive mechanical strength.

1.4.4 Curing of Epoxy Resin

Curing is a chemical reaction in which the epoxide groups of epoxy resin combine with a hardener to form a network with three dimensions. To be able to attain durability, impermeability, and stiffness, epoxy resins require the process of curing with a hardener. The curing process can proceed swiftly within a temperature range of 10°C to 150°C, depending on the curing method used.

1.4.5 Curing Agents

Depending on the procedure and desired qualities, a variety of curing agents can cure epoxy resins. Amines, polyamides, phenolic resins, anhydrides, isocyanates, and polymercaptans are the most often used epoxy curing agents. The structure of the hardener molecules determines how fast the system cures and what the glass transition temperature (Tg) is. Choosing the right resin and hardener relies on the application, manufacturing procedure, and desired output. The epoxy-hardener system's stoichiometry influences the finished material's physical characteristics. Using a variety of hardener types and quantities could have an impact on cross-link density and morphology. Epoxy resins tend to be cured utilizing amines and phenolic resin-based curing agents.

1.5 REINFORCEMENT

Reinforcement phases can be either continuous or discontinuous. It has excellent mechanical qualities. Whiskers, particles, or small strands with anisotropic properties comprise discontinuous reinforcement. Continuous reinforcement involves implanting anisotropic wires or fibers into the matrix in a specific direction. The kind, size, and form of reinforcement, along with the quantity and bonding between the matrix and reinforcement, primarily determine the composite behaviour.

1.5.1 Glass Fiber

Based on sales figures, glass fiber is the most commonly utilized material for reinforcement. After the invention of glass fiber, it was extensively used as an insulator and a material to strengthen composites. Glass fibers are classified into many types according to their composition and usage. E-glass is the predominant form of glass fiber utilized in applications involving composites because of its excellent mechanical and electrical insulating characteristics. S-glass is used in composites where there is a requirement for outstanding tensile strength. Nevertheless, this material is somewhat more expensive. Composite materials that require robust mechanical properties, such as fatigue life and chemical resistance, use R-glass. Glass fibers typically possess an average diameter ranging from 10 to 12 μm and a specific gravity of roughly 2.5.

1.5.2 Aramid Fibers

Aramid fibers are highly resistant to heat and strong synthetic materials. They have utility in aviation and military applications, specifically as fabric for body armour that meets ballistic standards and as an alternative for asbestos. This term refers to a truncated form of aromatic polyamide. This material has a high degree of molecular alignment along the direction of the fibers, capitalizing on the strength of the chemical bonds and demonstrating exceptional endurance.

1.5.3 Carbon Fiber

Carbon fiber, also referred to as graphite fiber, is a widely used form of fiber reinforcement in FRP. One of the most prevalent methods for producing carbon fiber is the oxidation and thermal combustion of polyacrylonitrile (PAN), often known as PAN-based carbon fibers. Carbon atoms make up this substance, which primarily consists of very thin threads with a diameter of roughly 5-10 um. The carbon atoms generate minute crystals that align parallel to the longitudinal axis of the fiber. This configuration results in the fiber exhibiting remarkably strong tensile properties. According to Kocoglu *et al.* (2023), carbon fiber possesses a tensile strength ranging from 3000 to 5800 MPa and an elastic modulus ranging from 500 to 600 GPa. Carbon fibers shows superior tensile strength and elastic modulus compared to glass fibers, despite their lower density. Carbon fiber's exceptional properties make it a perfect choice for enhancing composite materials in demanding applications, such as manufacturing aeroplane components, automotive parts, sporting equipment, and wind turbine blades.

1.5.4 Fabrics

Woven fabrics are made up of strands positioned perpendicular to each other. The angle of lay establishes the distinction between the warp and the weft. Looms produce woven fabrics in a wide range of weights, weaves, and widths, which tend to be stronger. Bi-directional woven fabrics provide strong strength in both the 0 and 90-degree orientations, as well as the capacity to use just one laminate for quicker composite production. Crimped fibers travel across and underneath each other to form these fabrics. Bi-directional fabrics contain many weaves. In a plain weave, every filler yarn or roving continuously passes over and beneath every warp fiber, as seen in Figure 1.5. Fabrics with a greater fabric count provide superior shape, strength, and durability. Woven fabric composites with reinforcement are becoming increasingly popular in the polymer composites industry because of their favourable fabric characteristics and simple production.

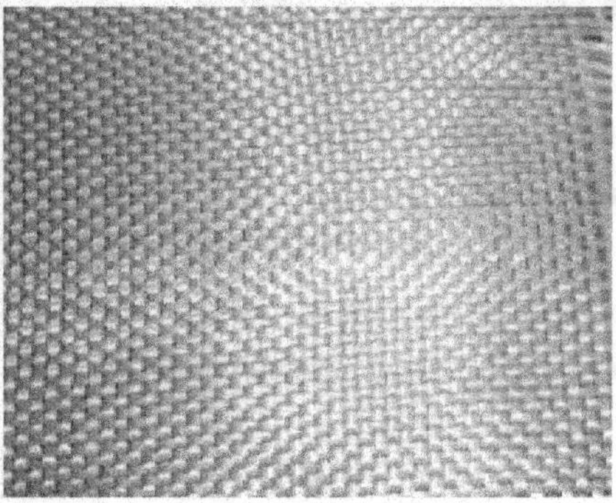

Figure 1.5 Glass fabric

1.6 MANUFACTURING OF COMPOSITES

Reinforced plastics are fabricated using a variety of methods, some of which are unique and others of which are based on typical polymer manufacturing techniques. The required blend, along with the scale and geometry of the developing structure, dictates the combination of fibers and matrix in a composite material. The composite is formed by blending the matrix and fibers, with the matrix impregnating and wetting the fibers. Several processes, including hand layup, bag moulding, filament winding, pultrusion, bulk moulding, compression moulding, sheet moulding, and resin transfer moulding, are available for processing thermosetting polymers.

1.6.1 Hand Layup Process

Hand layup is the most expensive method, but it requires minimal equipment. The automobile sector uses it for prototyping and low-volume manufacturing. During the layup process, one can perform hand layup using prepreg materials or dry cloth that has received resin reinforcement. Preparing prepregs into laminates with the required reinforcement and shape is a labour-intensive process that significantly contributes to the cost of high-performance composite constructions. Hand layup part fabrication accounts for more than 70% of the total cost of finished component manufacturing. Figure 1.6 shows the hand lay-up process's setup.

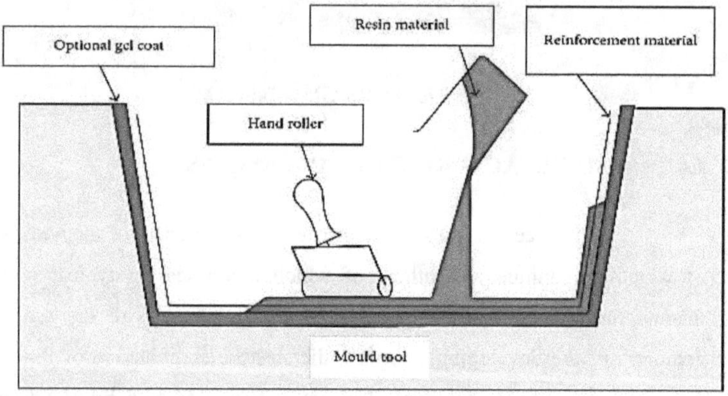

Figure 1.6 Set up of hand lay-up process

1.6.2 Filament Winding Process

Increasing the concentration and length of glass fiber reinforcement strengthens FRP products. The best structural performance is achieved by winding continuous glass roving that has been saturated with catalyzed resin around a spinning spindle. However, this technique is only applicable to goods with axial symmetry, like cylindrical products. Glass roving is wound using a bias angle to create a reciprocating helical pattern that meets structural requirements. The chemical industry commonly uses filament winding to fabricate pipes, tubes, cylinders, spheres, and massive tanks. Figure 1.7 displays the schematic of the filament winding process.

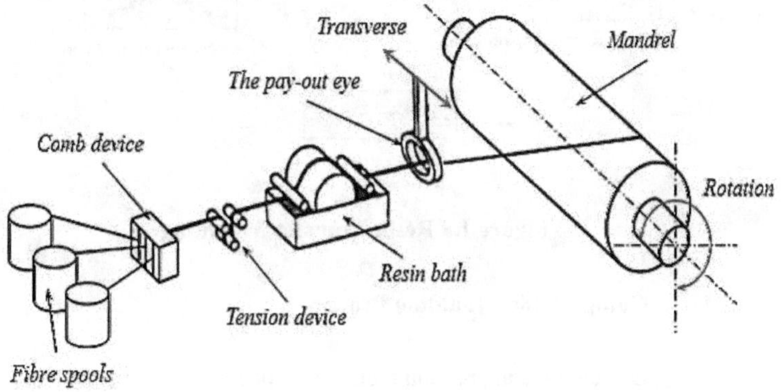

Figure 1.7 Schematic of Filament Winding process

1.6.3 Resin Transfer Moulding

Resin transfer moulding, as shown in Figure 1.8, allows for outstanding durability in composite products with unlimited geometry and rapid cycle times. The mould sets a dry reinforcing preform and closes it. The thermoset liquid resin impregnates the preform and then flows into the mould. During the injection process, the resin soaks the reinforcement and polymerises.

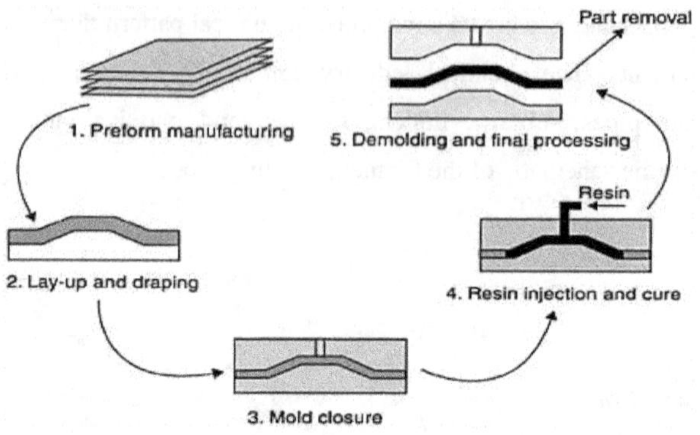

Figure 1.8 Resin Transfer Moulding

1.6.4 Compression Moulding Process

During the compression moulding process, the thermoset liquid will be contained within the mould and will adopt the intended shape. Subsequently, the mould will be subjected to heat in order to finalize the moulding procedure.

During the heating process, the liquid experiences many changes, such as the transition into a gel and the transition into a solid when subjected to pressure. After the completion of the previous step, the compression stress is exerted on the liquid while it is in the gel form. This enables the component to be moulded, as depicted in Figure 1.9. Compression loads determine the shape of the component, whereas heating induces the transformation of a liquid into a solid state.

Figure 1.9 Compression Moulding machine

1.7 TRIBOLOGY

Wear refers to the act of removing material from either or both solid surfaces that are in contact, typically when objects slide or roll against each other. The removal of materials may be gradual, but it is a constant process. The wear causes surface degradation and material loss from the contacting surface. Wear has a well-recognized impact on the reliability of industrial components, and its cost is also significant. Initial studies on wear began in various nations in the early 1960s.

Wear failures (i.e., part replacement) resulted in increased effort and time, manufacturing losses, downstream energy losses, and a greater environmental burden for enterprises. Despite substantial scientific research, wear concerns persist in industrial applications. Plastic deformation of surfaces and particle removal in the form of debris induce wear in metal, leading to a decrease in overall efficiency. The existence of wear can be detected in the following areas: metal-to-metal contacts or metal-to-nonmetal surface contacts. The two most frequent wearing processes in polymers are abrasive and adhesive wear.

Metal wear is induced by plastic deformation of the surface and particle separation in the form of wear debris, and it is one of the key causes contributing to a decline in system performance as a whole. Wear can occur in any of the following situations: metal-to-metal surface contact, flowing liquid particles, and nonmetals Panneerselvam *et al.* (2020). Adhesive and abrasive wear are the most noticeable characteristics in many polymer-based friction composites. Observation of the following wear phenomenon depends on the type of surface contact.

1.8 WEAR MECHANISM TYPES

1.8.1 Adhesive Wear

When two flat surfaces contact and glide over each other, there is a possibility of friction and material wear, which is known as an adhesive wear mechanism. This also provides a bonding force between the two surfaces, leading to loss of material across the sliding surfaces irrespective of the material hardness. In normal machine and equipment operation, a projected amount of wear occurs; however, too much friction produces earlier wear, which leads to equipment breakdowns, increased replacement expenses, equipment downtime, and decreased output by Vaisakh *et al.* (2016). The adhesive wear mechanism primarily depends on the following factors:

- Combined adhesive forces keep the interacting surfaces together.
- As demonstrated in Figure 1.10, the relative motion across the two contact surfaces, as well as the material's plastic

deformation, generates wear debris (hard particles) and the transfer of material from one surface to another.

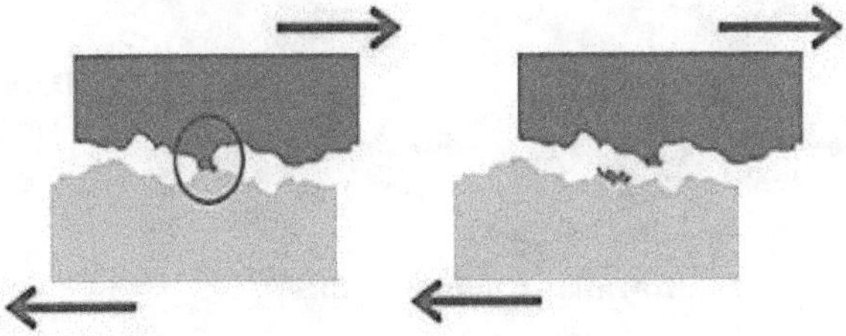

Figure 1.10 Adhesive Wear Mechanism

1.8.2 Abrasive Wear

Abrasive wear occurs when rigid protrusions from one object slide against a less rigid object under pressure, removing material from the softer object and creating a groove on its surface discussed by Gates *et al.* (1998). Abrasive wear occurs mainly by the removal of material by tough particles pressed on and moving along a solid surface, resulting in plastic flow and brittle fracture of the material system. This abrasive wear phenomenon is found in agriculture, mining, mineral processing, and rock cutting by Zum Gahr *et al.* (1998). Abrasive wear is one of the most common types, accounting for roughly 63% of overall wear costs. Abrasive wear can take two different forms: two-body abrasion and three-body abrasion by Suresh *et al.* (2009).

Figure 1.11 depicts a schematic model of abrasive wear with two and three bodies. The wearing mechanism in both two- and three-body abrasion is identical: the action of a comparatively hard material erodes the soft surface. When solid surfaces come into contact with strong protrusions, it leads to a type of wear called two-body abrasive wear.

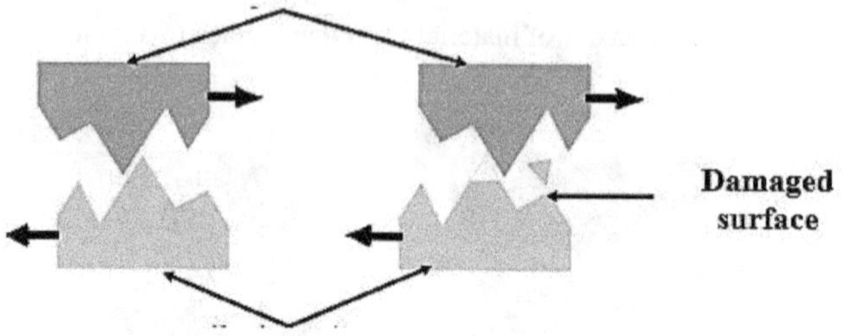

Figure 1.11 Abrasive Wear Mechanism

Two-body abrasive wear is commonly found in pipeline and chute liners used for transporting abrasive materials, as well as in blades and sections of agricultural and earth moving machines. It also occurs during crushing, cutting, and machining processes. Vehicle spring bushes, braking systems, clutches, chain wear strips, and rope bearings all experience three-body abrasive wear, as do sleeve bearings and bushes operating in abrasive environments.

1.8.3 Erosive Wear

Erosive wear is caused by the impact of solid, liquid, or gaseous particles that remove material from the surface due to the momentum effect. Figure 1.12 shows the erosive wear mechanism. Cavitation in the lubrication oil is responsible for the erosive wear of motor bearings. Cavitation voids, or bubbles, can form when the oil separates from the narrow space between the bearing and journal surfaces. The oil pressure rapidly decreases, resulting in the formation of empty spaces. These spaces disintegrate, creating a shock wave that eliminates particulates from the outermost layer.

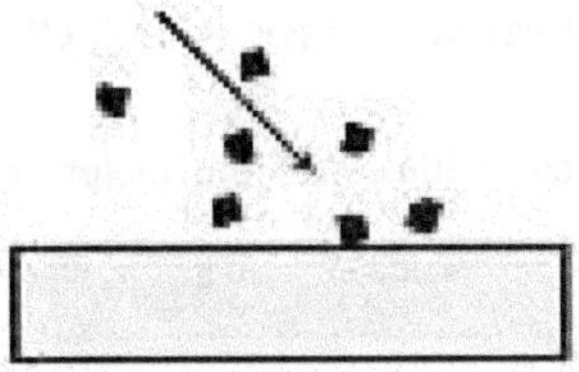

Figure 1.12 Erosive Wear Mechanism

1.8.4 Corrosive Wear

Corrosion, which happens when the surfaces in contact undergo oxidation, can intensify corrosive wear. Friction increases temperature and eliminates protective oxide coatings from the outermost surface, accelerating the oxidation process. Friction continuously deteriorates the oxide coating, leading to the formation of new oxide coatings. The three-body wear mechanism accelerates the rate of wear as the removal of hard oxide particles from the surface traps them between the sliding and rolling surfaces.

1.8.5 Fatigue Wear

Fatigue wear occurs as a result of repeated loading during friction. This phenomenon occurs when an external force exceeds the material's fatigue limit. Fatigue cracks originate at the substance's outermost layer and propagate into the material's subsurface regions. These fatigue-induced fissures may connect, leading to the separation of materials and subsequent delamination. Fatigue wear, often referred to as fretting wear, is the result of two surfaces sliding against each other in a repetitive way with a reduced range of motion. Fatigue wear occurs as a result of the surfaces' cyclic compression and tension. An engine bearing overlay serves as a notable example of fatigue wear, where cracks propagate through the intermediary layer until they completely disappear.

1.9 FACTORS DETERMINING THE WEAR OF MATERIALS

Figure 1.13 graphically illustrates the factors that determine the material's wear.

- The selection of materials for the matrix and filler in composite materials is crucial.
- Materials' characteristics include crystallinity, glass transition temperature, mechanical characteristics, molecular weight, orientation, toughness, and surface energy.
- The temperature, relative speed of contacting the surfaces, normal load, and surroundings all have an effect on polymer tribological properties. However, industrial application conditions vary widely, necessitating performance additives for each specific application scenario.

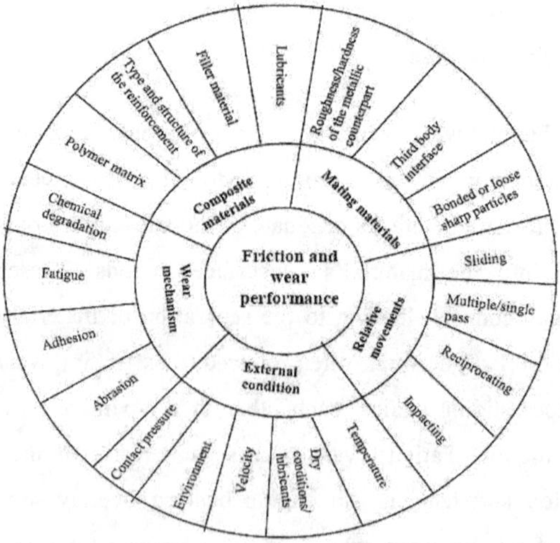

Figure 1.13 Factors affecting wear mechanism

1.10 TAGUCHI METHODOLOGY

Genichi Taguchi developed the Taguchi technique, which is a strong tool for developing high-performance, user-friendly systems. It introduces an organized approach to design for performance and quality optimization. Manufacturing companies and a variety of other industries frequently employ this method. By modifying design parameters, the Taguchi design can enhance efficiency and reduce quality variation. Minitab is a statistical software application that is frequently utilized to assess the effect of design elements. The Taguchi technique aids experiment performance by identifying optimal outcomes and key variables to consider. The Taguchi technique also utilizes orthogonal arrays, a novel structure, in conjunction with a limited number of trials to enhance the experiment's information gathering.

1.11 SIGNAL TO NOISE RATIO

The current experimental study then converts the results into a signal-to-noise (S/N) ratio. The term "signal" refers to the desired value (mean) of the output characteristics, whereas "noise" refers to the unwanted value. As a result, the S/N ratio is equal to the square deviation divided by the mean. The S/N ratio in analogue and digital communications measures signal intensity in relation to background noise, often expressed in decibels. Taguchi suggests employing the S/N ratio to determine whether quality characteristics deviate from desired levels.

The S/N analysis calculates the S/N ratio for each level of process parameters. A higher S/N ratio corresponds to superior quality attributes, regardless of category. A statistical ANOVA is used to determine which process factors are statistically significant and which play a significant role. The S/N ratio and ANOVA analysis can be utilized for determining the optimal combination of process parameters. Typically, the analysis of the S/N ratio identifies two broad types of quality features: the lower the better and the higher the better.

1.12 SUMMARY

This chapter discusses the introduction of composites, composite material classification, polymer composites, epoxy materials, types of reinforcements, composite manufacturing, tribology, and Taguchi methodology.

CHAPTER 2

LITERATURE SURVEY

2.1 INTRODUCTION

This chapter provides an extensive literature survey about manufacturing of composites, characterization of composites made by hand layup and filament winding process, mechanical characterization of composites, tribological characterization, and Taguchi techniques. This chapter concludes with a list of research gaps.

2.2 MECHANICAL CHARACTERIZATION

Perumal *et al.* (2024) did a study that looked at the mechanical and physical features of epoxy polymer composites that were strengthened with glass fibers and changed with rutile nanoparticles. The addition of rutile nanoparticles to the epoxy matrix resulted in substantial mechanical improvements. The evenly distributed nanoparticles throughout the matrix improved the tensile strength, flexural strength, and impact resistance. This led to better bonding between the fibers and epoxy at the interface. The morphological properties of the composites changed when rutile nanoparticles were added. This made the fibers and nanoparticles spread out more evenly throughout the matrix. As a result, there was an improvement in fracture toughness and the ability to resist crack propagation. The study suggests that adding rutile nanoparticles to glass fiber-reinforced epoxy composites could lead to the creation of new materials with tailored mechanical properties.

Turaka *et al.* (2023) study investigated the enhancement of mechanical characteristics in glass/epoxy composites using a hybrid blend of multi-walled carbon nanotubes (MWCNTs) and graphene nanoparticles. Using a sonication technique, the researchers created the hybrid filler and evenly distributed it within the epoxy resin matrix. The composites produced by the authors showed increased tensile, flexural, and impact strength when compared to both pure epoxy and composites containing only one type of filler. The use of the hybrid filler system improved the process of transferring and distributing loads more effectively in the matrix, resulting in better mechanical characteristics. Furthermore, the combined effects of MWCNTs and graphene nanoparticles played a role in improving the performance. This study demonstrates the ability of hybrid filler systems to improve the mechanical characteristics of polymer composites, opening up possibilities for their use in a variety of engineering domains.

Perumal *et al.* (2023) examined the effects of adding zircon particles to glass fiber-reinforced epoxy composites. The investigators evaluated both mechanical and morphological features in order to gain knowledge about the performance of the composite material. The study discovered that the inclusion of zircon particles significantly improved the composite's mechanical durability, indicating potential use in industries that require resilient materials. Additionally, the morphological study demonstrated a better distribution and stronger connection between the zircon particles and the epoxy matrix, indicating an improvement in structural strength. The results emphasize the potential of zircon particles in strengthening composite materials, leading to progress in material science and engineering. The research offers useful insights for creating high-performance composites with customised features, hence creating opportunities for various industrial applications.

Nur *et al.* (2022) conducted research on improving the mechanical properties of E-glass fiber-reinforced epoxy resins by reinforcing with fillers. They conduct experiments to study how different types and concentrations of fillers can improve the mechanical properties of these composites. The study aims to enhance tensile strength, flexural strength, and impact resistance by changing the compositions of fillers. They evaluate the enhancements in performance that fillers achieve through thorough mechanical testing and analysis. The results show significant improvements in the mechanical properties, suggesting that fillers have the ability to reinforce epoxy resin composites. This study offers valuable insights into the advancement of high-performance materials for various engineering applications, such as structural components in construction, automotive parts, and aerospace components.

Rajhi *et al.* (2022) fortified the composites with a blend of several nanoparticles to improve their mechanical characteristics. The addition of nano-fillers led to improvements in tensile strength, flexural rigidity, and impact resistance when compared to the equivalents without fillers. In addition, the combination of different types of nano fillers resulted in synergistic effects, which further improved the mechanical properties of the composites. The study also examined the impact of nanofiller concentration on the mechanical properties, offering valuable information for tailoring the composite composition for specific applications.

Sapiai *et al.* (2020) investigated the tensile and flexural properties of unidirectional kenaf and hybrid glass/kenaf epoxy composites reinforced with silica nanoparticles. The study's goal was to improve the mechanical properties of kenaf fiber-reinforced epoxy composites using nanoparticles. Their research revealed substantial enhancements in both tensile and bending characteristics with the incorporation of silica nanoparticles. In addition, the combination of kenaf with glass fibers led to synergistic improvements, demonstrating increased strength in comparison to composites made solely from kenaf. This study highlights the potential of silica nanoparticles in enhancing the mechanical properties of natural fiber-reinforced epoxy composites.

Megahed *et al.* (2019) examined how the addition of silica and carbon nanoparticles affects the mechanical properties of hybrid glass fiber-reinforced epoxy composites. The investigators reported that including silica nanoparticles in the composites led to increased tensile and flexural strength. The researchers attributed this improvement to the nanoparticles' greater interfacial adhesion and dispersion inside the epoxy matrix. In addition, the inclusion of carbon nanoparticles resulted in a further enhancement of the mechanical properties. This improvement can be attributed to the combined effects of silica and carbon nanoparticles working together. The hybrid composites demonstrated superior toughness, impact resistance, and fatigue performance in comparison to both the neat epoxy and the glass fiber-reinforced epoxy composites.

Aigbodion *et al.* (2017) investigated the mechanical properties of a composite material, a blend of glass fibers and epoxy, reinforced with silica nanoparticles. The researchers also examined the impact of incorporating silicon nanoparticles on the performance of the composite material. Their study suggests that the use of silica nanoparticles resulted in enhancements in mechanical characteristics, including heightened tensile and flexural strength, as well as improved hardness and impact resistance. The hybrid composite showed outstanding properties in comparison to conventional materials, suggesting its potential for diverse industrial applications. This study elucidates the capacity of silica nanoparticles to enhance the mechanical characteristics of composite materials, therefore propelling advancements in the realms of tribology and industrial applications.

Kalaiselvam *et al.* (2014) conducted the research to examine how the addition of nanofillers affects the thermal and mechanical properties of glass fiber-reinforced epoxy composites. The authors discovered that the use of nanofillers improves both the mechanical strength and thermal stability of the composites. More precisely, the inclusion of nanofillers improves characteristics such as flexural strength, tensile strength, and resistance to impact, indicating better overall performance of the combined material. Moreover, the incorporation of nanofillers significantly improves the thermal properties, including thermal conductivity and heat deflection temperature. The results show that nanofillers have the potential to enhance the mechanical and thermal properties of glass fiber-reinforced epoxy composites.

Hinojos-Mendoza *et al.* (2014) investigated the mechanical properties of epoxy composites reinforced with short glass fibers and alumina microparticles. The study aimed to assess the effect of including alumina microparticles when combined with glass fibers on the mechanical properties of composites. The study investigated factors such as strength, stiffness, and toughness using multiple testing methods, including tensile and flexural tests. The results showed that the inclusion of alumina microparticles resulted in improvements in mechanical characteristics, such as increased tensile and bending strength, along with higher toughness. This research shows that short glass fibers and alumina microparticles can work together to make epoxy composites with better mechanical properties. This means that these composites can be used in many different fields.

Dorigato *et al.* (2013) studied whether hybrid organic-inorganic nanofillers affect the mechanical characteristics and fracture toughness of epoxy composites strengthened with glass fibers. The purpose of the investigation was to assess how the addition of these nanofillers affected the strength, stiffness, and fracture resistance of the composite material. Adding hybrid organic-inorganic nanofillers to the materials in the study made the mechanical properties much better, as shown by tests like tensile testing and fracture toughness tests. More precisely, the composites demonstrated increased tensile strength, higher rigidity, and improved resistance to fracture in comparison to base materials. The results make a point of showing that hybrid organic-inorganic nanofillers can make glass fiber-reinforced epoxy composites stronger.

Asi *et al.* (2009) examined the mechanical characteristics of epoxy composites reinforced with glass fibers and filled with Al_2O_3 particles. The study sought to investigate the impact of including Al_2O_3 particles on the mechanical properties of the composite. Experimental study revealed that the addition of Al_2O_3 particles resulted in improvements in several mechanical parameters, including tensile strength, flexural strength, and impact strength. The addition of Al_2O_3 particles to the composite resulted in increased stiffness and resistance to deformation. This suggests that filler reinforcement has the potential to improve the performance of glass-fiber reinforced epoxy composites. The findings improve the understanding of composite material design and propose practical applications in areas that require enhanced mechanical qualities.

2.3 DYNAMIC MECHANICAL ANALYSIS

Raja Othman *et al.* (2022) carried out studies on an innovative model that explores the combined impacts of hybrid micro and nanosilica on the mechanical characteristics of epoxy composites. The inclusion of these particles in the epoxy matrix results in improved composite strength, rigidity, and resilience. Their discoveries offer a hopeful pathway for creating epoxy composites with customized mechanical characteristics for a wide range of uses, including structural elements and innovative engineering materials. This study adds to the growing knowledge of nanocomposites materials and their ability to improve material performance in a variety of domains.

Bommegowda *et al.* (2021) study examined how a combination of micro- and nanofiller hybrids influences the dynamic mechanical properties of glass-reinforced epoxy composites. The goal is to enhance the mechanical properties of these composites by integrating fillers at both micro and nanosizes. By performing an order of experiments and analysis, they evaluate factors such as storage modulus, loss modulus, and damping qualities. The study investigates the impact of different filler compositions and concentrations on the dynamic mechanical properties of the composites. The results show that the use of filler hybrids significantly improves the mechanical properties, particularly in terms of rigidity and vibration absorption capacity. This research provides vital insights for enhancing the design and performance of epoxy composites used in applications that need exceptional mechanical qualities.

Kuruvilla *et al.* (2020) conducted a study to investigate the mechanical properties of glass fiber-reinforced epoxy composites enhanced with both micro and nanofillers. Their primary goal was to understand the impact of including these fillers on the composite material's mechanical properties. As part of their study, they looked for any changes in mechanical properties, like damping, storage modulus, and dynamic mechanical behaviour, which might be due to the addition of micro and nanofillers. The study utilized DMA to evaluate how the material reacted to different frequencies and temperatures. The findings from this study contribute to the advancement of composite materials with customized mechanical characteristics, which could potentially have a positive impact on several industries.

Yamunadevi *et al.* (2019) in their study, investigated the characteristics of epoxy nanocomposites produced by combining woven roven glass fiber with cerium-zirconium oxide. They conducted a dynamic mechanical analysis and examined the mechanical characteristics of the manufactured composites. Adding cerium-zirconium oxide nanoparticles to the epoxy matrix improves the mechanical properties. They also concluded that there was an improvement in stiffness and damping characteristics of the manufactured composites.

Karthicksundar *et al.* (2016) assessed the thermal properties of carbon nanoparticle-reinforced GFRP. They used a dynamic mechanical analyzer to evaluate the material's response at different temperatures. The objective of incorporating carbon nanoparticles into the GFRP matrix was to improve its heat conductivity and stability. By conducting trials, they noticed alterations in the dynamic mechanical properties of the material, such as its storage modulus and damping characteristics, in reaction to changes in temperature. The findings revealed that the carbon nanoparticle-infused GFRP had a higher thermal efficiency than traditional GFRP materials.

Ferreira Capela *et al.* (2011) examined the mechanical properties of composites made from a combination of fiber and glass microspheres, focusing on their dynamic behavior. The researchers conducted dynamic mechanical analysis to investigate the influence of glass microspheres on the mechanical characteristics of fiber-reinforced composites. The study investigates the behaviour of composite materials when subjected to dynamic loading conditions in order to understand their reaction to different stress levels over a period of time. They evaluate variables such as rigidity, resistance to motion, and the ability to deform and return to its original shape through experiments. The findings provide insight into the synergistic benefits of including various reinforcement materials, demonstrating possible enhancements in mechanical performance in comparison to conventional composites composed solely of fibers. Gaining a comprehensive understanding of these dynamic features is essential for maximizing the efficiency and effectiveness of composite materials in many engineering fields, ranging from automotive to aerospace sectors.

2.4 FATIGUE CHARACTERISTICS

Goumghar *et al.* (2022) investigated the fatigue behaviour of hybrid flax-glass/epoxy composites. The research aimed to explore how combining flax and glass fibers in an epoxy matrix influenced the material's fatigue resistance. They evaluated the composite's performance under cyclic loading conditions using experimental analysis. The study gave new information about how the hybrid composite wears down over time. It showed how things like fiber orientation, matrix properties, and interface interactions affect how it behaves mechanically. The findings provide valuable information for the design and optimisation of hybrid fiber composites for applications requiring durability and reliability under cyclic loading.

In their study, Fathy *et al.* (2017) looked at how adding nanosilica and nanoalumina changes the fatigue performance of epoxy composites reinforced with glass fibers. They concluded that incorporating these nanoparticles improved the material's ability to withstand fatigue. The researchers attributed this improvement to the strengthening of the link between different interfaces and the reduction in crack spread. The researchers determined that including nanosilica/nano-alumina into the composites significantly improved their durability and lifetime when subjected to cyclic loading conditions.

Borrego *et al.* (2014) investigated experimentally the fatigue properties of epoxy composites reinforced with glass fibers and different types of nanofillers. The main aim of this study was to investigate the impact of various nanofillers on the fatigue characteristics of the manufactured composites. The experiment's results showed that the addition of nano-fillers improved fatigue strength and crack propagation resistance when compared to composites without fillers. Furthermore, they noted changes in fatigue behaviour based on the specific nanofiller employed, indicating that different fillers may have distinct impacts on the properties of the composite material. The results highlight the capability of nano-fillers to significantly improve the fatigue properties of glass fiber-reinforced epoxy composites.

Manjunatha *et al.* (2010) studied the fatigue properties of epoxy composites reinforced with glass fibers and different types of nano-fillers. The main aim of this study was to investigate the impact of various nanofillers on the fatigue characteristics of the manufactured composites. The experiment's results showed that the addition of nano-fillers improved fatigue strength and crack propagation resistance when compared to composites without fillers. Furthermore, they noted changes in fatigue behavior based on the specific nanofiller employed, indicating that different fillers may have distinct impacts on the properties of the composite material. The results highlight the capability of nano-fillers to significantly improve the fatigue properties of glass fiber-reinforced epoxy composites.

2.5 HEAT DEFLECTION TEST

Zhou *et al.* (2022) examine the effect of adding wollastonite filler on the thermal stability of epoxy composites reinforced with glass fibers. The study looks into what happens to the thermal stability of manufactured materials when different amounts of wollastonite filler are added to the epoxy matrix of glass fiber composites. The findings show that adding wollastonite filler boosts the thermal stability of the composites, probably because it promotes the bonding between the glass fibers and epoxy matrix at the interface. The improvement in performance is due to the presence of wollastonite filler, which acts as a reinforcing agent. This reduces the risk of thermal damage and slows the onset of decomposition at high temperatures.

Wang *et al.* (2019) looked at how adding carbon nanotube (CNT) fillers changed the HDT of glass fiber-reinforced epoxy composites. The HDT was investigated by adding various filler compositions to the epoxy matrix of glass fiber composites. The findings indicate that adding CNT fillers improves the composite material's heat resistance by increasing its thermal conductivity. This reinforcement serves to prevent deformation when exposed to elevated temperatures. The precise concentration of CNT fillers and their distribution within the composite material affect the extent of improvement observed in the HDT. The study suggests that the addition of CNT fillers to glass fiber-reinforced epoxy composites enhances their thermal stability, rendering them appropriate for high-temperature applications.

Li *et al.* (2019) assessed the thermal stability and degradation rate of glass fiber/epoxy composites that include expanded graphite fillers. Using thermal analysis and calorimetry, the study investigated the effects of including expanded graphite on the composite material's thermal stability and degrading behaviour. The findings demonstrate that the inclusion of expanded graphite enhances the thermal stability of the composites, which delays the process of degradation. The study offers valuable insights into how the addition of expanded graphite fillers boosts the thermal performance of glass fiber/epoxy composites. This could have major implications for a variety of industrial uses that require materials having enhanced resistance to heat.

Lyon *et al.* (2005) investigated the flammability of glass fiber-reinforced polymer composites. They analyzed the flammability of the materials through testing and analysis. Their findings provide information about the flammability and potential hazards associated with glass fiber-reinforced polymer composites, which is critical for industries that rely on such materials. This research contributes to enhancing safety measures and informing material selection for applications in fire-prone environments.

2.6 TRIBOLOGICAL PROPERTIES

Vishwakarma *et al.* (2021) conducted a study to examine the friction and wear characteristics of glass fiber-reinforced epoxy composites containing various additives when subjected to sliding without lubrication. Researchers conducted experiments to evaluate friction and wear, observing notable impacts on both characteristics due to the presence of various fillers. Several fillers showed potential for mitigating friction and wear, indicating the possibility of improving the performance of composites in industrial applications. Understanding these behaviors is critical for companies that require materials with superior tribological qualities, including the aerospace and automotive industries.

Ragab *et al.* (2021) explored how the inclusion of nanoclay enhanced the friction and wear properties of glass fiber-reinforced polyurethane nanocomposites. The study aims to improve the material's friction and wear characteristics for various industrial applications. Testing revealed that the inclusion of nanoclay significantly enhanced the tribological performance of the manufactured materials. The improvement has been attributed to the increased distribution of nanoparticles within the polymer matrix, resulting in reduced wear and friction rates. The results of this research emphasize the possibility of nanoclay as a potential additive to enhance the friction and wear properties of composite materials, providing valuable information for the creation of stronger and more efficient engineered materials.

Wang *et al.* (2021) looked at how friction and wear affected epoxy composites that had glass fibers and functionalized graphene oxide (FGO) added to them. They implemented the incorporation of FGO in the composite to investigate its capacity to improve friction and wear characteristics. Testing revealed that the inclusion of FGO significantly improved the tribological qualities. The composite demonstrated reduced friction coefficients and better wear resistance in comparison to samples without FGO. This study emphasizes the potential of FGO as a filler in glass fiber-reinforced epoxy composites, providing valuable information for the creation of innovative materials with improved tribological properties for an extensive variety of uses in industry.

El-Tayeb *et al.* (2020) conducted a study that looked at the effects of graphene nanoplatelets (GNPs) on the friction and wear parameters of epoxy/glass fiber composites. The researchers discovered that the addition of GNPs improved the composites' friction and wear properties. The unique lubricating and strengthening characteristics of graphene led to an overall decrease in friction and wear rates with the inclusion of GNPs. The even distribution of GNPs inside the epoxy matrix enhanced the ability of the composite material to withstand loads and prevent surface damage, thus enhancing its overall durability. This study suggests that adding graphene nanoplatelets to epoxy-based composite materials might be a good way to improve their resistance to friction and wear. This has major implications for engineering applications that require greater wear resistance and better frictional behaviour.

Padil *et al.* (2020) investigated the application of green-synthesized epoxy-glass fiber nanocomposites for tribological purposes. The composites were made using an environmentally friendly method, and their characteristics were extensively investigated. The study aimed to comprehend the mechanical properties of the nanocomposites. The results showed that the epoxy glass fiber nanocomposites, which were made using environmentally friendly methods, had good mechanical properties that made them a good choice for tribological applications. In this study, the use of an environmentally conscious synthesis process proposes a viable strategy for the sustainable manufacturing of composite materials. This study advances the development of eco-friendly composite materials with favourable mechanical properties, providing potential opportunities for their use in various thermodynamic environments.

Mazid *et al.* (2020) conducted a study on techniques to improve the wear and friction properties of E-glass/epoxy composites. Their main objective was to integrate graphene and graphene oxide nanoplatelets into the composite material. They observed substantial enhancements in the composites' tribological characteristics after conducting thorough testing and evaluation. The incorporation of graphene and graphene oxide nanoplatelets resulted in lower friction and wear rates, demonstrating improved durability and performance during sliding conditions. The researchers attributed these enhancements to the reinforcing and lubrication properties of the nanoplatelets included in the combined matrix. Their research indicates potential opportunities for creating new composite materials that possess exceptional tribological properties. A wide range of industries that demand high-performance substances for friction and wear-resistant components might use these materials.

Jindal *et al.* (2020) conducted an investigation into the tribological properties of hybrid glass fiber epoxy composites reinforced with both nano-SiC and micro-SiC fillers. They conducted experimental tests to analyze the composites' friction coefficient and wear rate while changing filler concentrations. The findings showed that the inclusion of both nano-SiC and micro-SiC fillers resulted in a significant decrease in the composites' coefficient of friction and rate of wear in comparison to the composites without the fillers. The study attributed the enhancement in performance to the combinatorial impact of the two fillers, which strengthened the interfacial adhesion between glass fibers and the epoxy matrix. The research shows that hybrid composites with nano-SiC and micro-SiC fillers have good tribological properties. This means that they could be used in situations where they need to be very durable against wear.

Basak *et al.* (2019) investigated the influence of wollastonite on the mechanical and tribological properties of GFRP composites. The researchers conducted a study to investigate the impact of incorporating wollastonite into the composite to enhance its performance. Tests revealed that the GFRP composite's mechanical and tribological properties improved with the addition of wollastonite. In particular, the composite material demonstrated improved strength and resilience to wear in comparison to materials that did not contain wollastonite. This study shows that wollastonite could be used as a strengthening agent in GFRP composites. This information is useful for making new materials that are better at withstanding mechanical and tribological forces and can be used in industry.

Kumar *et al.* (2019) investigated the mechanical and tribological properties of glass fiber-epoxy composites that were infused with nano-ZrO_2 particles. They performed tests to evaluate how different amounts of nano-ZrO_2 affected the characteristics of the composites. The study revealed that the inclusion of nano-ZrO_2 resulted in improvements in the mechanical and tribological properties of the composites. These enhancements include improved hardness, tensile strength, and wear resistance. In addition, the study highlighted the possibility of nano-ZrO_2 as a favourable substance for enhancing the characteristics of glass fiber-epoxy composites. This provides essential knowledge for the development of composite products with improved performance and longevity.

Selvamani *et al.* (2019) investigated the impact of nanoclay on the friction and wear properties of epoxy composites reinforced with glass fibers. Their research showed that adding nanoclay significantly improved the composites' tribological performance. The inclusion of nanoclay resulted in decreased friction and wear rates, suggesting improved mechanical characteristics. The findings show that nanoclay has the potential to be a beneficial additive for improving the efficiency of composite materials.

Wu *et al.* (2018) investigated the effect of nano-SiO_2 on the tribological characteristics of glass fiber/epoxy composites. Wu *et al.* (2018) conducted tests to assess the friction coefficient and wear rate of composites containing varying amounts of nano-SiO_2. The results showed that adding Nano-SiO2 significantly reduced the friction coefficient and wear rate of the composites compared to the base material without Nano-SiO_2. The inclusion of nano-SiO_2 particles facilitated improved adhesion between the epoxy matrix and glass fibers, thereby enhancing performance. The study suggests that the addition of Nano-SiO_2 to glass fiber/epoxy composites can significantly improve their ability to withstand friction and wear, making them highly suitable for situations requiring strong resistance.

Gul *et al.* (2018) conducted a study and investigated the tribological characteristics of E-glass fiber-reinforced polypropylene composites, which include different fillers. Their goal was to gain in-depth knowledge of how different fillers influence the friction and wear properties of composites through a comprehensive tribological study. The researchers ran trials in controlled circumstances to evaluate the coefficient of friction and wear rate. Their research revealed that the inclusion of fillers had a significant impact on the composites' tribological properties. Certain additions showed improvements in reducing friction and abrasion, consequently enhancing the overall effectiveness of the materials.

Günaydin *et al.* (2018) conducted an investigation into the tribological characteristics of epoxy composites impregnated with glass microspheres. The investigators carried out research to evaluate the impact of incorporating glass microspheres on the wear and friction properties of the epoxy composites. Their research showed that the inclusion of glass microspheres resulted in enhancements in tribological characteristics, such as the reduction of friction and wear rates. These improvements have the potential to provide benefits in other fields, such as engineering and production, where reducing friction and wear is essential. The findings improve our understanding of the influence of various filler materials on the performance of composite materials, providing valuable insights for future studies and potential practical applications in industries that require materials with enhanced tribological capabilities.

Kalia *et al.* (2017) investigated the friction and wear properties of glass Fiber-reinforced epoxy composites. The properties were improved by adding nano-SiO_2 and nano-Al_2O_3 particles. The researchers aimed to improve the composites' resistance to wear by incorporating these nanoparticles. Under various conditions, the investigators conducted tests to investigate the friction and wear properties of the manufactured composite materials. The results showed that the inclusion of nano-SiO_2 and nano-Al_2O_3 particles led to better tribological performance in comparison to the base material. More precisely, the composites showed decreased friction and wear rates, indicating possible uses in industries that need materials with enhanced durability and resistance to wear.

2.7 TAGUCHI DESIGN OF EXPERIMENTS

Kechagias *et al.* (2024) presented effective methods for integrating the potential of experimental design into stochastic problems. The f-ME 3D printing technology and ABS filament was chosen for this study because of the numerous processing parameters that influence the quality and flexural strength of 3D printed objects. Consequently, researcher's determined optimum values for infill density, raster orientation, printing speed, layer thickness, and nozzle and bed temperatures using Taguchi orthogonal array and general linear (ANOVA).

Pathak *et al.* (2024) carried out a study with the aim of optimizing the machining parameters for glass fiber-reinforced textile composites. Their objective is to determine the optimal settings for attaining desired machining outcomes, such as high-quality surface finish and precise dimensions, by systematic experimentation and analysis. The study examines various milling parameters, such as cutting speed, feed rate, and depth of cut, to determine their impact on the composite material's machinability. Their goal is to improve process efficiency and productivity while avoiding tool wear and material damage using a systematic optimization technique. The study's findings offer valuable insights into the machining characteristics of glass fiber-reinforced textile composites.

Sharma *et al.* (2023) concentrated on developing and evaluating the frictional properties of epoxy composites reinforced with E-glass fibers and reinforced with fly ash. They use the Taguchi approach for their experimental evaluation. The study seeks to optimize the material composition of these composites to improve their tribological features, such as wear resistance and frictional behaviour. They conduct systematic experiments and analyses to examine how several factors, such as the concentration of filler, speed of sliding, and amount of applied force, impact the composites' tribological performance. They effectively identify the most beneficial combination of factors for achieving enhanced tribological properties by applying the Taguchi approach. The study's findings offer valuable insights into the synthesis and optimization of composite materials for thermodynamic applications. These insights can benefit the automotive, aerospace, and manufacturing industries, where friction and wear resistance are crucial factors.

Sharma *et al.* (2022), used Taguchi's Design of Experiments (DOE) and artificial neural network (ANN) analysis to guess how well GNP filled glass fiber-reinforced epoxy composites would work in dry sliding conditions. This research intends to optimize the composition of these composites in order to improve their tribological qualities, which include their resistance to wear and their behaviour when subjected to friction. In order to examine the influence of numerous parameters on the tribological performance of the composites, they conducted systematic experiments and used artificial neural network modeling. Some of the parameters that they investigated include GNP concentration, sliding speed, and applied stress. It is possible for them to accurately predict and optimize the tribological behaviour of the composites through the integration of Taguchi's DOE with ANN analysis.

Kumar *et al.* (2015) utilized Taguchi's technique to conduct a tribological analysis of nanoclay/epoxy/glass fiber composites. The research's objective was to improve the composites' tribological characteristics by utilizing Taguchi's experimental design method to manipulate variables such as load, sliding distance, and sliding speed. The findings exposed the impact that these parameters had on the composite's friction and wear properties, providing valuable insights into its behavior when subjected to sliding conditions. Taguchi's methodology facilitated methodical experimentation, which in turn enabled the discovery of optimal parameter combinations for the purpose of improving tribological performance.

2.8 FILAMENT WINDING PROCESS

Liang *et al.* (2023) looked into how the curing temperature affected the mechanical properties of carbon fiber composites. They focused on how well these composites could be used to build filament winding pressure vessels. They conducted a series of experiments where they changed the curing temperatures in order to evaluate how these changes affect the tensile, compressive, and flexural properties of the composite materials. The study investigates the impact of various curing temperatures on the matrix curing process, adhesion between fibers and matrix, and the composite's overall performance. Outcomes offer useful insights into refining the manufacturing process for carbon fiber composite pressure vessels by analyzing the relationship between mechanical test results and variations in curing temperature. Their discoveries help to improve the design and efficiency of filament-wound pressure vessels, which are critical to various sectors such as aerospace, automotive, and renewable energy.

Smith *et al.* (2020) focused on the process of selecting and characterizing resins for filament winding of glass fiber composites. This is a crucial factor that affects the mechanical characteristics and overall performance of finished products. The study methodically assesses several resin systems, taking into account characteristics such as viscosity, curing kinetics, and compatibility with glass fibers. The study aims to identify resin compositions that provide the most desirable processing features and improve the mechanical characteristics of the resultant composites through experimental analysis and characterization approaches, such as rheological measurements and thermal analysis. The study establishes a correlation between the qualities of resin and the specifications of filament winding, as well as the performance of the resulting composite. This correlation offers useful insights into the process of selecting and optimizing resin systems for filament winding applications. This study improves the understanding of resin behaviour in composite material manufacturing processes, thereby aiding in the production of excellent filament-wound glass fiber composites for a variety of industrial uses.

Minsch *et al.* (2017) conducted a study where they analyzed filament winding techniques and prospective equipment technologies. The study examines the optimization of filament winding variables such as tension, speed, fiber orientation, and resin choice in order to improve the properties of composite materials. Their research covers cutting-edge equipment technologies such as automated systems, process monitoring, and quality control procedures. The study seeks to enhance the efficiency, precision, and dependability of filament winding techniques used in diverse applications in industry. The output provides valuable insights into how to improve the production of high-quality composite structures by evaluating these technologies.

Lee *et al.* (2017) have published a review article. The review article provides a thorough examination of filament winding techniques, the materials employed, the process parameters, and the applications across different sectors. The authors explore the benefits of filament winding in the production of outstanding-performance composite structures and emphasizes current progress in this area. The authors thoroughly examined the process characteristics and mechanical aspects of glass fiber filament winding. Their study investigated the multiple aspects that impact filament winding, including winding tension, speed, fiber orientation, and selection of resin. It emphasizes the importance of adjusting these parameters to improve mechanical properties like tensile strength, flexural strength, and interlaminar shear strength. The review provides useful information for the production of excellent composite structures by analyzing the complex link between process factors and the material's final performance. This synthesis enhances the comprehension and improvement of glass fiber filament winding procedures, which are vital for companies that depend on composite materials for structural purposes.

Wang *et al.* (2017) examined how the addition of microscale fillers might improve the impact resistance of filament-wound glass fiber composites. This work specifically examines the mechanical characteristics of composite materials that are essential for a range of engineering uses. The objective was to enhance impact resistance by including microscale fillers into the composite matrix. By doing experimental analysis and testing, they proved that the addition of these fillers had a beneficial impact on the composite's ability to endure influences. The study provides valuable insights into the design and optimization of composite materials to improve their mechanical properties.

Wang *et al.* (2017) investigated how microscale fillers affect the abrasion resistance of filament-wound glass fiber composites. They conducted a study on the influence of adding these fillers on the composite's resistance to abrasive forces, with a specific focus on materials science and engineering. Researchers noticed that the addition of microscale fillers resulted in enhanced composite abrasion resistance through experimental analysis and testing. The outcomes indicated a possible approach to improve the strength and lifespan of composite materials in several fields where the capacity to withstand abrasion is important, such as in marine, automotive, and building sectors. The study provides vital insights into optimizing composite materials, which could lead to breakthroughs in developing more durable and enduring components for engineering applications.

Garcia *et al.* (2017) looked into what happens to the mechanical properties of filament-wound glass fiber composites when nanoclay fillers are added, both when the loads are static and when they are moving. Their research aimed to understand the impact of including nanoclay fillers on the composite's behavior under different loading conditions. The researchers conducted a thorough experimental investigation, which involved both static and dynamic mechanical testing. As a result, they found changes in the stiffness, strength, and energy absorption capabilities of the composite material. These results showed that adding nanoclay fillers improved the mechanical properties of the materials when they were loaded in both static and dynamic ways. Applications in aerospace, automobiles, and structural engineering can benefit from this study's useful insights into enhancing the mechanical properties of composite materials under various loading circumstances.

Li *et al.* (2016) examined the mechanical and electrical properties of hybrid filament-wound glass fibers containing CNT fillers. The research investigated the synergistic impact of using novel CNT fillers in filament winding methods in conjunction with standard glass fibers. Experimental analysis and testing were used by the researchers to assess the mechanical strength and electrical conductivity of the composite materials. Their research revealed enhancements in both mechanical and electrical characteristics when compared to traditional glass fiber composites. The use of CNT fillers enhanced the composite's tensile strength and modulus, while also increasing its electrical conductivity. This study provides significant contributions to the advancement of composite materials with multifunctional properties, which have potential uses in several sectors.

Stefanovska *et al.* (2015) investigated the bending properties of composite pipes through theoretical modelling and experimental testing. Their study aims to assess bending stiffness and strength, crucial for various engineering applications. The researchers developed predictive equations based on material and geometric factors and validated them through experiments. They better understand how composite pipes behave under bending loads by comparing theoretical predictions with experimental results. This leads to better design and optimization processes for these kinds of structures.

Zhang *et al.* (2008) introduce a framework for modelling and simulating the process of glass fiber filament winding. The focus of this work is on developing computational methods to predict the behavior of fibers and resin during the winding process. The model accurately replicates the process of composite structure production by taking into account variables like tension, winding speed, and fiber placement. They also verified the validity of their model by comparing it to experimental data, showcasing its precision in forecasting filament winding results. This study contributes to the optimization of process parameters, the enhancement of production efficiency, and the improvement of filament-wound composite materials' mechanical characteristics. As a result, it advances the knowledge and utilization of filament winding in various industries.

Smith *et al.* (2001) conducted a study to investigate the complex glass fiber filament winding processes specifically designed for aviation applications. The study explores novel techniques to improve the effectiveness and productivity of filament winding procedures in the aviation industry's composite manufacturing. Smith and his colleagues explore subjects including sophisticated filament winding machinery, innovative fiber implantation techniques, and enhanced resin formulations. The need for establishing accurate fiber alignment and regulated resin impregnation is highlighted to guarantee the appropriate mechanical characteristics and structural integrity of aerospace components. The paper explores the progress made in automation, process monitoring, and quality control procedures to fulfil the demanding aerospace standards for lightweight, strong materials. The study addresses important technical difficulties, which contribute to the ongoing growth of new composite materials and production methods that are critical for use in the aviation industry.

2.9 PRESSURE TESTING

Martinez *et al.* (2021) conducted a study on the evaluation of leaks and the strength of filament-wound glass fiber composites with rubber-based fillers under high-pressure conditions. They sought to evaluate the structural soundness and impermeability of these composites under high-pressure conditions. The researchers studied the efficacy of the composite materials by conducting a thorough evaluation, which involved leakage testing plus integrity assessment. Their research provides important insights into the suitability of using filament-wound glass fiber composites with rubber-based fillers in high-pressure-resistant applications, such as pressure vessel technology-related fields.

Williams *et al.* (2021) conducted pressure fatigue tests on filament-wound glass fiber composites reinforced with hybrid fillers. The study's goal was to assess the durability and resistance to fatigue of these composites tested under cyclic loads. Their research provides valuable insights into the fatigue characteristics and long-term durability of filament-wound glass fiber composites reinforced with hybrid fillers. Experiments provide valuable insights into pressure loads, which are prevalent in various engineering fields.

2.10 DROP WEIGHT IMPACT TESTING

Rodriguez *et al.* (2021) conducted drop-weight impact tests on filament-wound glass fiber composites with microscale fillers, specifically for aeronautical applications. The study's goal was to assess the impact resistance and performance of these composites when tested under dynamic loading conditions that are typical of aircraft environments. The composite's capacity to withstand abrupt impact loads was evaluated using drop-weight impact testing techniques in an experimental analysis. The findings of this study provided significant insights into the behavior and structural integrity of filament-wound glass fiber composites with microscale fillers. These findings are critical for ensuring the security and dependability of aircraft components that may encounter impact events.

Smith *et al.* (2017) conducted drop-weight impact tests on glass fiber composites enhanced with nanofillers, made using a filament-winding process. The study's goal was to assess the impact resistance and performance of these composites when exposed to dynamic loading conditions, which frequently occur in structural applications. The researchers evaluated the composite's capacity to endure abrupt impact loads by conducting experimental studies using drop-weight testing methodologies. The researchers gained significant insights into the behavior and strength of filament-wound glass fiber composites reinforced with nanofillers. The current study deepens our understanding of nanofiller-strengthened composite materials, offering guidance for the development of high-performance materials capable of withstanding impact loads.

2.11 RESEARCH GAP

Based on an extensive literature survey, the present study has identified the following research gaps.

- Several researchers have reinforced glass fiber in an epoxy matrix with various fillers, but no one else has reported the addition of zinc borate.
- No researchers has reported manufacturing of epoxy/glass fiber/zinc borate composite pipes through filament winding process using zinc borate as filler.
- The previous review of the literature showed that there weren't many studies on how to process and characterize epoxy/glass fiber/zinc borate composites in terms of their shape, strength, and temperature.
- No one else has reported on the tribological test involving abrasive wear of epoxy, glass fiber, and zinc borate composites.

- There were also no studies on the processing and characterization of epoxy/glass fiber/zinc borate (morphological, mechanical, hydrostatic pressure, vibration, and water absorption) by the filament winding process.

CHAPTER 3

MATERIALS AND METHODS

3.1 INTRODUCTION

This chapter deals with the objectives of the present research, methodology of the entire research, materials selected, manufacturing process involved, material characterization, and testing procedures.

3.2 OBJECTIVES OF THE RESEARCH

- To fabricate composites containing Glass Fiber/Epoxy/Zinc borate materials by the compression moulding process.
- To characterize Glass Fiber/Epoxy/Zinc borate composite material by means of morphological, mechanical and thermal methods of characterization.
- To assess the friction and wear characteristics of Glass Fiber/Epoxy/Zinc borate composites by two body abrasive wear process.
- To analyze the abraded surfaces by means of scanning electron microscope.
- To fabricate composite pipe structures containing Glass Fiber/Epoxy/Zinc borate materials using filament winding process.

- To test the Glass Fiber/Epoxy/Zinc borate composite pipes by means of compression, drop weight impact, hydrostatic pressure, vibration and water absorption.

3.3 METHODOLOGY

The flow diagram of the present research work is shown in the Figure 3.1.

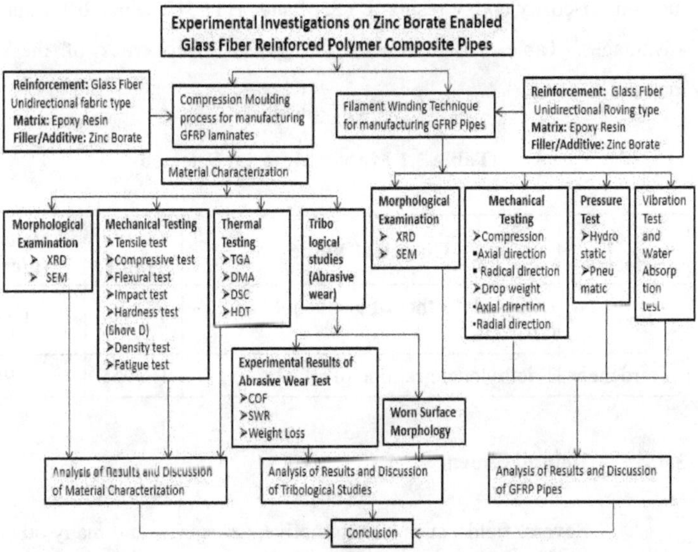

Figure 3.1 Methodology of the research

3.4 SELECTION OF MATERIAL

3.4.1 Matrix

Epoxies are highly valued in the thermoset engineering polymers category for their exceptional mechanical qualities and resistance to chemicals and corrosion. Moulded or cast epoxies provide exceptional dimensional stability and minimal shrinkage. Engineering often uses the DGEBA, an epoxy compound. It has two epoxide groups, one at each end of the molecule. The curing or hardening reaction triggers the conversion of resin from a liquid state to a solid state. The hydrogen atoms in the curing agents react chemically with epoxide groups at room temperature or with hanging hydroxyl groups at higher temperatures. This process establishes a network of interconnected links. This present research investigation employs the medium-viscosity epoxy resins 'LY556' and 'HY951', which offer numerous advantages. Table 3.1 represents the physical properties of the matrix material.

Table 3.1 Matrix Material Selected

Material Selected	Chemical Name	Grade Name	Density (g/cm^2)
Epoxy	Diglycidyl Ether of bisphenol-A (DGEBA)	LY556	1.15
Hardener	Triethylenetetramine (TETA)	HY951	0.96

3.4.2 Reinforcement

Several fields, such as automotive, aerospace, and many others, use glass fibers because of their characteristics such as high impact strength, corrosion resistance, light weight, and thermal expansion. As illustrated in Figure 3.2, the present research study used E-glass in the form of fabrics and filament as a reinforcement.

Figure 3.2 Unidirectional Glass fiber and Fabric

3.4.3 Zinc Borate

Zinc borate, an inorganic compound, functions as a useful filler in epoxy matrices, thus enhancing numerous properties of the material. Initially, it provides smoke suppression and flame retardancy, which are essential for protecting applications that are susceptible to fire. The inclusion of zinc borate also improves mechanical properties, including tensile strength, hardness, and wear resistance. However, these improvements are contingent upon the particulate size and dispersion within the matrix. Proper dispersion of zinc borate is important for good performance. This is usually done with dispersing agents, ultrasonic dispersion, or high-shear blending. Various industries, including electronics, construction, automotive, and marine, employ zinc borate-filled epoxy resins to enhance fire safety, thermal resistance, and durability. The current study typically used 50 nm-sized zinc borate nanoparticles (ZBP) as fillers, sourced from Sigma Aldrich, as illustrated in the Figure 3.3. ZBP melts at 980°C and has a density of 3.64 g/cm^3.

Figure 3.3 Zinc Borate

3.5 ULTRASONICATION

ZBP was incorporated into the epoxy matrix through the ultrasonication procedure to ensure uniform dispersion within the epoxy resin. Sonication was a frequently employed technique in the past to ensure a uniform dispersion of the nanofiller in the matrix resin. A glass beaker was used to measure and add the appropriate quantity of nano ZBP (5 wt.%, 10 wt.%, and 15 wt.%) to the epoxy resin. The ingredients were combined using a stirrer. An instrument emitting 20-kHz sound waves further agitated the mixture. Sonication was conducted at ambient temperature for 30 minutes. A nanogel solution was made by adding a TETA hardener (HY951) after the nanogel was thoroughly mixed.

3.6 FABRICATION OF COMPOSITE LAMINATES

A compression moulding process was employed to create composite layers. Initially, the woven glass fiber mat was cut into six 300 mm × 300 mm squares to produce a laminate thickness of 3 mm that was suitable for a variety of mechanical testing activities. Employed a precise and uniform quantity of nanogel solution to fabricate composite laminates. The aluminium foil was initially treated with a releasing agent, followed by the nanogel solution and the glass fiber substrate. Applied an additional layer of nanogel solution to the glass fiber base. Similar to the preceding phase, the five remaining glass fiber mats were stacked one on top of the other, with a nanogel solvent coating between the layers. A final coating of nanogel solution was applied to the sixth glass fiber mat. Figure 3.4 illustrates the process of manufacturing the composite laminates. A stainless steel mould measuring 300 mm in square and 20 mm in thickness equips the compression moulding machine. To achieve a consistent thickness, the composite laminates were compressed in a mould press for one hour at 80 °C and 2.844 kPa. The composite material was post-cured at 60°C for an hour at the same pressure before being removed from the mould as the final product. Furthermore, as illustrated in Table 3.2, an unaltered variant of GFRP composite was manufactured without zinc borate.

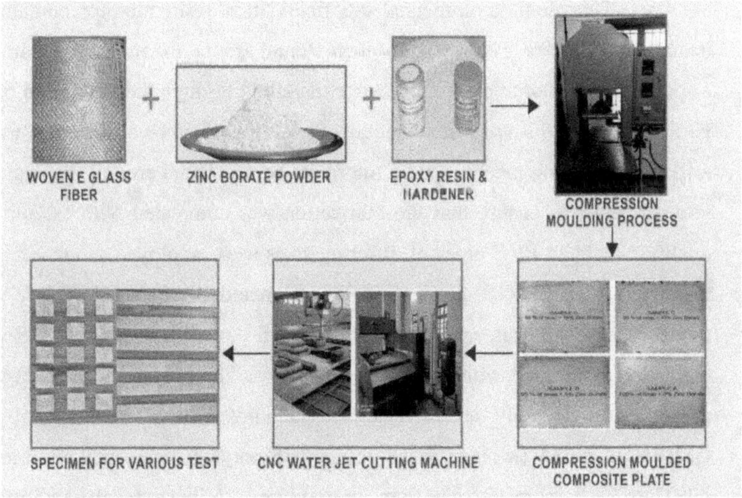

Figure 3.4 Fabrication of Glass Fiber/Zinc Borate Composites

Table 3.2 Composition of manufactured Glass Fiber/Zinc borate composite and its coding

Composite composition by wt. %	Composite coding
50 % Glass Fiber + 50% Epoxy resin	GF00
50% Glass Fiber + 5 wt. % Zinc Borate + 45% Epoxy resin	GF05
50% Glass Fiber + 10 wt. % Zinc Borate + 40% Epoxy resin	GF10
50% Glass Fiber + 15 wt. % Zinc Borate + 35 % Epoxy resin	GF15
GF	Glass Fiber
ER	Epoxy resin
ZBP	Zinc Borate Particles

3.7 MANUFACTURING OF GFRP PIPES

This method immerses the fibers in a resin mixture containing nanogel ZBP fillers. The fibers are then wound around the turning mandrel in a specific pattern or at a 45° angle after travelling through the ZBP resin bath. Nevertheless, in this specific technique, the entire process was executed using a filament winding machine, with the fibers being wound around the mandrel at a 45° angle to ensure that the fabrication was completed with the utmost quality. Steel and PVC pipes of differing sizes were employed as mandrels to manufacture GFRP/ZBP pipes. Initially, the mandrel's external surface was treated with mould oil. Subsequently, the axial direction of the mandrel was meticulously aligned with polyethylene strands. The polyethylene sheet was subsequently securely attached to the mandrel. The de-molding of the GFRP/ZBP pipe was made much easier with polyethylene strips and mould oil. The GFRP composite was then cured. Figure 3.5 illustrates the addition of zinc borate nanoparticles to epoxy to enhance its strength, as well as the filament winding process used to fabricate composite pipelines.

Figure 3.5 GFRP Pipes made using Filament winding machine

Figure 3.6 Fabricated samples of GFRP composites with 0, 5, 10, and 15 % ZNB

As depicted in Figure 3.6, the filament winding procedure constructed each layer of the composite pipe samples with a winding angle of approximately 45 degrees, resulting in a total of six layers. Each pipe's interior diameter is 59 mm, and the wall thickness is 2 mm. The impregnated pipe was extracted from the mandrel once the desired width was attained. The next step involved a curing period of 150 minutes at 100°C.

3.8 X-RAY DIFFRACTION ANALYSIS

Using XRD, polymer or composite samples can be examined to obtain important information regarding their solid-state structure. For XRD analysis, the scanning angle range of 10°–30° was chosen. Bragg's equation (nλ=2dsinΘ) was used to calculate the d-spacing. An XRD analysis was performed on every composite that was produced.

3.9 DENSITY AND VOID FRACTION

The ratio of void volume to total volume using the rule of mixture was established by the Broutman and Agarwal method as shown in equation (3.1) were effective to calculate the theoretical density of composites.

$$\delta_{ce} = \frac{1}{\frac{w_f}{\delta_f} + \frac{w_m}{\delta_m}} \tag{3.1}$$

Where w and δ represent the weight fraction and constituent density, respectively. The composite material was denoted by the suffix ce, the fiber was denoted by the suffix f, and the matrix was denoted by the suffix m. As the zinc borate filled GFRP composite consisted of a matrix, Fibers, and particle filler, the equation (3.1) may be rewritten as equation (3.2) with some minor adjustments.

$$\delta_{ct} = \frac{1}{\frac{w_f}{\delta_f} + \frac{w_m}{\delta_m} + \frac{w_p}{\delta_p}} \tag{3.2}$$

Whereas, the suffix p indicates the particulates filler components being used. The Archimedes principle was used to arrive at an accurate value of the composite's density. Using equation (3.3), were effective to determine the void volume fraction (V_v) of the composite.

$$V_v = \frac{\delta_{ct} - \delta_{ce}}{\delta_{ct}} \times 100 \tag{3.3}$$

3.10 HARDNESS TEST

In accordance with the ASTM D2240-05 requirements, we conducted the hardness test for the manufactured composites using a Shore D hardness tester, as shown in Figure 3.7 (b). The Figure 3.7 (a) shows the samples used for testing.

Figure 3.7 (a) GFRP sample for hardness test

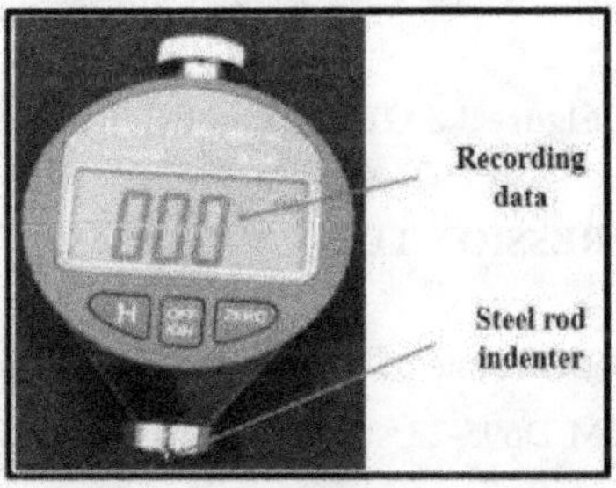

Figure 3.7 (b) Shore D hardness tester

3.11 TENSILE TEST

The tensile tests were conducted on the produced composites using a computerized tensometer following ASTM D3039. Figure 3.8 depicts the macrograph of the tensile test specimen. The specimen size is a rectangle with a width of 25 mm X length of 250 mm X thickness of 3 mm, and the tensile tests were conducted at a 1 mm/min crosshead speed.

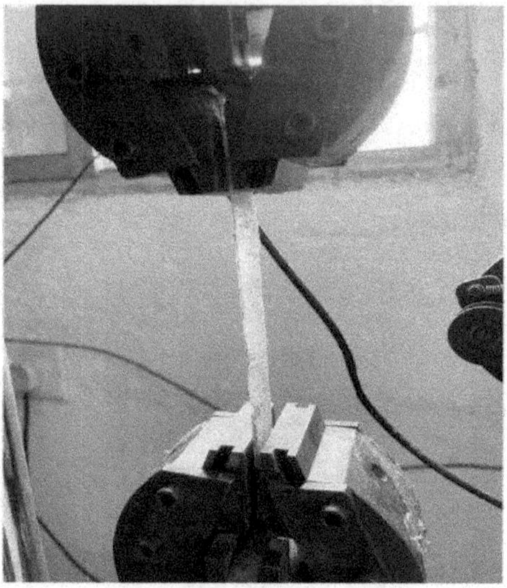

Figure 3.8 GFRP sample for tensile test

3.12 COMPRESSION TEST

Compression tests of the manufactured composites were evaluated according to ASTM D695-23 standard test method which was developed by ASTM international for measuring compressive properties of rigid plastics, including both unreinforced and reinforced materials such as high – modulus composites. The GFRP samples are considered as rigid plastics and experiment was done. The specimen size is 12.7 mm width X 100 mm length X 3 mm thickness. The strain rate was maintained at 2 mm/min. The compression test specimen is shown in Figure 3.9.

Figure 3.9 GFRP sample for Compression test

3.13 FLEXURAL TEST

The flexural tests were conducted on fabricated composites using the three-point bending mode per the ASTM D790. Figure 3.10 displays the flexural specimen's macrograph. The specimen size is 3.2 mm thickness X 12.7 mm width X length of 127 mm. Flexural testing was conducted at a crosshead speed of 2 mm/min.

Figure 3.10 GFRP sample for Flexural test

3.14 IMPACT TEST

The impact tests were conducted on the manufactured composites utilizing a computerized impact testing machine following ASTM D256, and the dimensions of the impact test specimens are depicted in Figure 3.11.

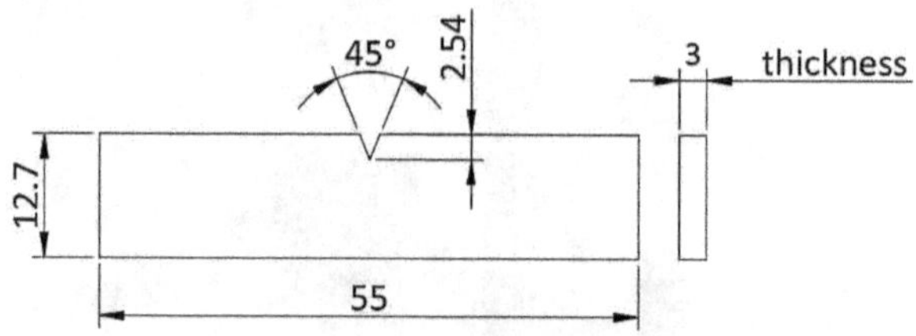

Figure 3.11 Test sample geometry for Impact test

3.15 DIFFERENTIAL SCANNING CALORIMETRY

The manufactured composites were analyzed on a DSC 6000-Perkin Elmer apparatus. Under a nitrogen atmosphere, the scanning range was set from 30°C to 220°C. The DSC test used a 10°C/min heating and cooling rate in order to find the melting point (T_m), the crystalline point (T_c), the specific melting enthalpy (H_f), and the degree of crystallinity (X_c).

3.16 THERMO GRAVIMETRIC ANALYSIS

TGA is a method used to identify specific characteristics of materials that experience a decrease in mass as a result of decomposition, oxidation, or the release of volatile substances. The TGA sample was heated at 10°C/min, at nitrogen atmosphere and the temperature of the sample was increased at constant rate. The temperature range was 30^0 C to 600^0 C. The thermal stability of the manufactured composites was determined using the TGA 4000 Perkin Elmer instrument by TGA analysis. A sample weighing 5-10 mg of manufactured composite materials was used for TGA analysis.

3.17 WEAR TEST

In order to determine the frictional and wear values, it is imperative to examine the tribological properties of polymers and polymer composite materials. Abrasive wear experiments were conducted on the manufactured composites in accordance with the ASTM G99-05 using a DUCOM TR-20LE pin-on-disc tribo-tester, as illustrated in Figure 3.12, to investigate their tribological properties. The sample size is 10 mm diameter X 30 mm height.

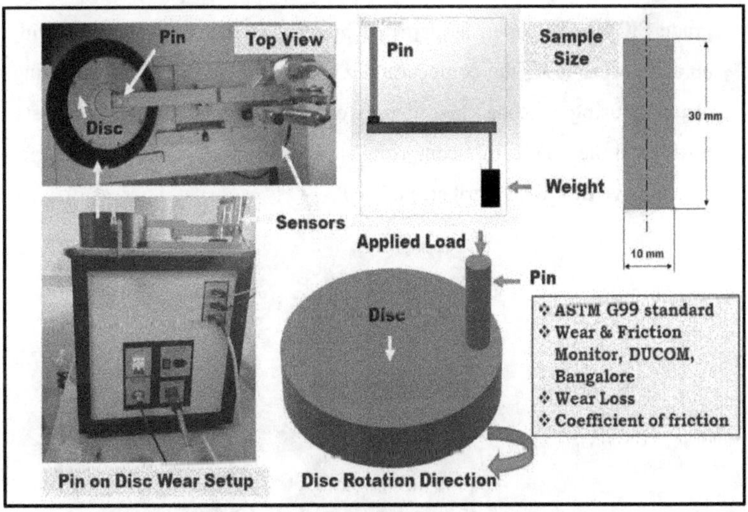

Figure 3.12 Pictorial Representation of a pin-on-disc-tribo-tester

The rotational speed and test duration are calculated using the equations (3.4) and (3.5)

$$\text{Sliding velocity (m/s)} = \pi DN / 60{,}000 \quad (3.4)$$

$$\text{Sliding distance (m)} = \pi DNT / 60{,}000 \quad (3.5)$$

where, D - Track diameter (mm), N - Rotational Speed (rpm) and T - Test duration (s).

The control panel utilizes the angular velocity and the duration of the test as inputs. The loading unit is exposed to the required load, and the experiments are carried out. In this tribological investigation, a rotating disc made of AISI D3 steel was coated with 320 grit abrasive paper and utilized as a counter face. In order to achieve full contact between the pin and abrasive disc surface, the composite pin specimens were initially pre-worn using 600-grade SiC emery paper, as depicted in Figure 3.13. Prior to conducting the abrasive wear tests, the composite pin specimen was subjected to drying and cleaning using acetone. In order to carry out the abrasive wear tests, we positioned the HDPE/PBI composite sample at a right angle to the surface of the abrasive disc with a grit size of 320.

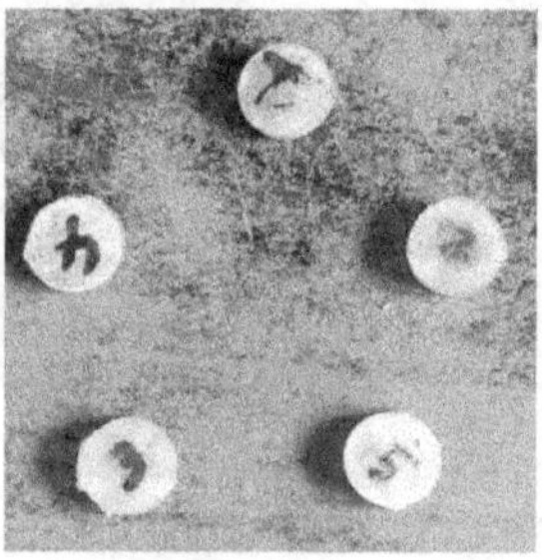

Figure 3.13 GFRP sample for wear test

The input parameters for the abrasive wear test selected were (i) percentage of filler (wt. %) (ii) sliding distance (m) and applied load (N) and the abrasive wear output parameters measured was (i) weight loss, (ii) CoF and (iii) SWR. The abrasive wear tests were carried out at a sliding distance of 0.5 (m/s). The initial weight before and after the abrasive wear tests were both weighted using an electronic digital analytical scale with a

0.0001 g reading for the purpose of determining the amount of weight loss experienced by the specimen. Equation 3.6 was used to compute the particular wear rate.

$$\text{Specific wear rate} = \frac{m_1 - m_2}{\rho \times N \times S} \quad (3.6)$$

Where m_1 and m_2 - Mass of the pin specimen before and after the experiment (g)

ρ - Density of the specimen (g/mm^3),

N - Applied load (N),

S - Sliding distance (m).

Abrasive wear experiments were conducted with frictional force being measured directly and CoF computed using Equation (3.7).

$$\text{Coeffecient of friction} = \frac{\text{Frictional Force (F)}}{\text{Applied Load (N)}} \quad (3.7)$$

3.18 HEAT DEFLECTION TEMPERATURE

HDT is a crucial factor in the design of plastic products as it quantifies their ability to withstand high temperatures. The HDT tests were conducted by the ASTM D648. The manufactured composite samples were placed beneath the deflection measuring instrument and subjected to a load of 0.45 MPa. The manufactured composites were immersed in a silicone oil bath and the temperature was raised at a rate of 2°C per minute. The distortion temperature was determined by recording the temperature of the heat transfer medium when the test specimen deflected by 0.25 mm.

3.19 DYNAMIC MECHANICAL ANALYSIS

DMA is a technique employed to examine the viscoelastic properties of polymers and polymer composites. The use of DMA can provide valuable insights into the frequency and temperature-dependent properties of a solid-state composite material. The increasing use of composites in dynamic environments such as cars, industrial settings, and structural applications necessitates an examination of their dynamic load-bearing properties. In this investigation, the DMA characterization of the produced composites was conducted using specimens measuring 50 × 10 × 4 mm. The fabricated composites underwent a heating process from 20°C to 150°C, with a heating rate of 10°C/min. This process was carried out under a load of 1000 N and a frequency of 15 Hz. Factors obtained from DMA tests are storage and loss modulus. Figure 3.14 displays the samples for the Dynamic mechanical analysis test.

Figure 3.14 GFRP sample for Dynamic mechanical analysis test

3.20 FATIGUE TEST

Fatigue studies are used to examine the durability or lifetime of a material when it is subjected to cyclic loads. The experiment was conducted in accordance with ASTM D7774 utilizing fatigue testing equipment (Zwick/Roell, LTM) equipped with a 2 kN load cell, operating in a 3-point bending mode. Due to the time-consuming nature of the analysis, all the samples underwent 2×10^5 cycles of loading. The study is conducted in force controller mode to determine the potential deformation under cyclic flexural stress. The samples underwent testing at 50% of the maximum flexural load, with a frequency of 5 Hz and an amplitude of 5N. The deformation during each cycle was documented to examine its durability characteristics. The experimental setup of the fatigue testing machine is shown in the Figure 3.15(a), and the fatigue test specimen is shown in the Figure 3.15 (b).

Figure 3.15 (a) Fatigue test setup

Figure 3.15 (b) GFRP sample for Fatigue test

3.21 COMPRESSION BEHAVIOR FOR GFRP COMPOSITE PIPE

Compression tests were performed using an INSTRON Universal testing machine with a maximum capacity of 500 kN, in accordance with the ASTM D695-15. This study was conducted to test filament wound cylinders. The test was conducted using an aspect ratio of 2:1, where the diameter was twice the length. Tests were conducted both axially and in the radial direction, as shown in Figure 3.16 (a) and Figure 3.16 (b).

Figure 3.16 (a) Axial Compression Setup

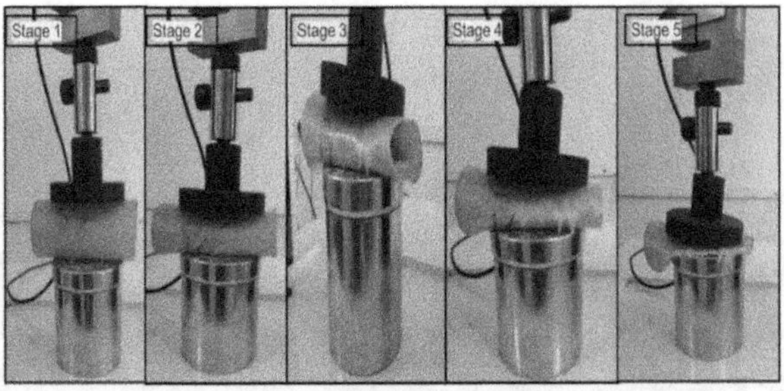

Figure 3.16 (b) Radial Compression setup

3.22 DROP WEIGHT IMPACT TESTING FOR GFRP COMPOSITE PIPE

Drop-weight impact testing is conducted according to ASTM D7136. It assesses the ability of the GFRP composite pipe to withstand impact events caused by dropping weights. This test assesses the durability of GFRP composite pipes by measuring their resistance to concentrated impacts and identifying the specific type of damage caused. The test is conducted in two different orientations based on the positioning of the specimen in the Low velocity drop weight machine: axial direction and radial direction, as shown in Figure 3.17. An experiment was conducted to analyze the effects of repeatedly dropping a load from a certain height.

3.23 HYDROSTATIC PRESSURE TEST FOR GFRP COMPOSITE PIPE

The ASTM D1599 was employed to perform internal pressure testing on the composite pipes that were manufactured. Figure 3.18 depicts the test setup. Every experiment consisted of filling a sample with oil that has a low compressibility and water, making it suited for hydraulic applications.

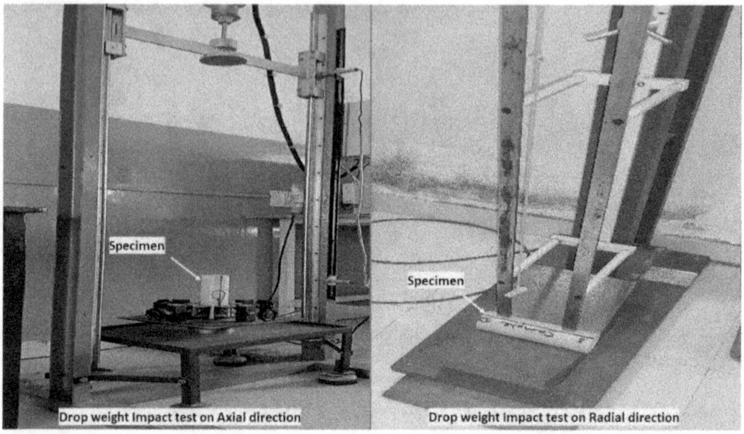

Figure 3.17 Drop Weight Impact test on Axial Direction and Radial Direction

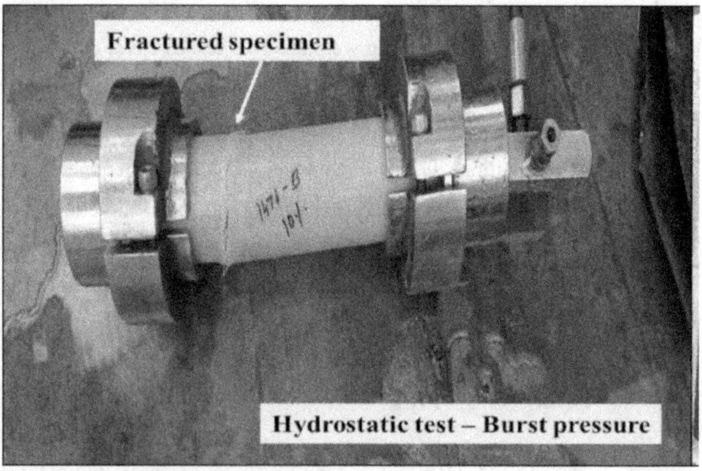

Figure 3.18 Hydrostatic pressure test setup

3.24 PNEUMATIC PRESSURE TEST FOR GFRP COMPOSITE PIPE

Figure 3.19 illustrates the arrangement of GFRP pipes, which are equipped with a valve at one end and an end cap at the other end. The compressed pneumatic air or gas is introduced into one end of the GFRP pipe and allowed to fill the interior. The pressure is maintained at a steady level for a duration of 10 minutes. This generates a high-pressure atmosphere within the pipe, allowing for an assessment of the pipe's dependability, performance, safety, and ability to prevent leaks in the pressure system prior to its implementation in real-world scenarios.

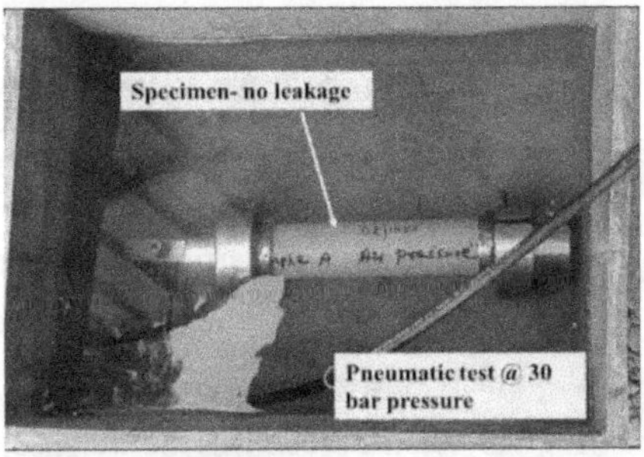

Figure 3.19 Pneumatic pressure test setup

3.25 VIBRATION TEST FOR GFRP COMPOSITE PIPE

Vibration is one of the most common causes of internal surface damage in pipe systems. To assess the impact of vibration based on natural frequencies, three end conditions, namely one end free, one end fixed, and both ends free, were applied to the fabricated GFRP pipes. An experiment similar to modal analysis as shown in Figure 3.20 is carried out for glass fiber-reinforced polymer pipes with three distinct places. The natural frequencies are obtained, and an experimental verification of the impact of wall-thinning on the pipe's modal frequencies was made.

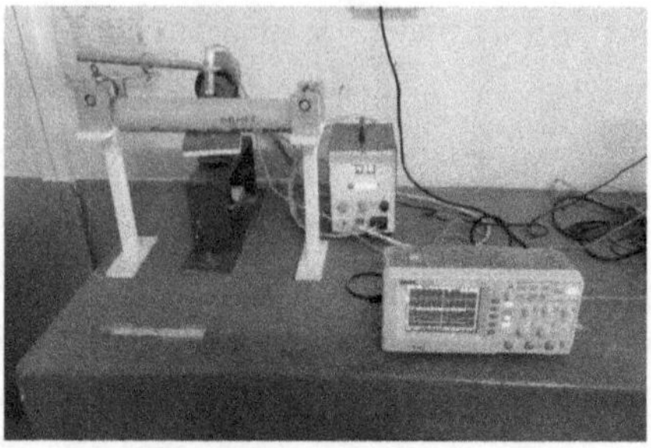

Figure 3.20 Vibration test setup

3.26 WATER ABSORPTION TEST FOR GFRP COMPOSITE PIPE

A water absorption test was carried out to regulate the amount of water to be absorbed by the specimen. For this experiment, ASTM standard D570 was selected. Initially, the specimens were dried in the oven to remove the moisture content. Further, the specimens are weighed, and they are dipped into various types of water, like distilled water, normal water, and salt water. The material is then emerged in water at agreed-upon conditions, often 23°C. These experiments were taken at a time interval of 48 hours. After 48 hours, the specimens were taken out, cleaned properly, and weighed. Then calculate the weight difference of these specimens in terms of how much percentage was increased. The procedure can be repeated for 48 hours for all the specimens under 3 different water environments.

3.27 SUMMARY

In this chapter, a summary of the research objectives, a flow diagram of the research, the materials that were chosen, the fabrication procedures that were involved, and the testing methodologies were all described.

CHAPTER 4

MATERIAL CHARACTERIZATION

4.1 PREAMBLE

This chapter will provide an in-depth discussion on the various experimental material characterization procedures that are carried out for the composite specimens manufactured through the compression moulding technique. There is no investigation on the possibility of using zinc borate as a filler, despite the fact that there are numerous fillers that are reinforced into the glass fiber. Using a compression moulding process, the purpose of this study was to generate a novel form of GFRP composite that was reinforced with ZBP at weight percentages ranging from 0 to 15 percent. A comprehensive analysis of the morphological, mechanical, thermal, and physical properties of manufactured composites was carried out.

4.2 X-RAY DIFFRACTION AND MORPHOLOGICAL STUDY OF THE COMPOSITES

The presence of zinc borate nanoparticles appeared to have an effect on the crystallographic behavior of glass fiber. In order to establish the existence of the glass fiber and zinc borate composites, as well as the bonding of the interlayer distance and phase structure, an indirect method known as XRD was utilized. Figure 4.1 depicts the same thing as well.

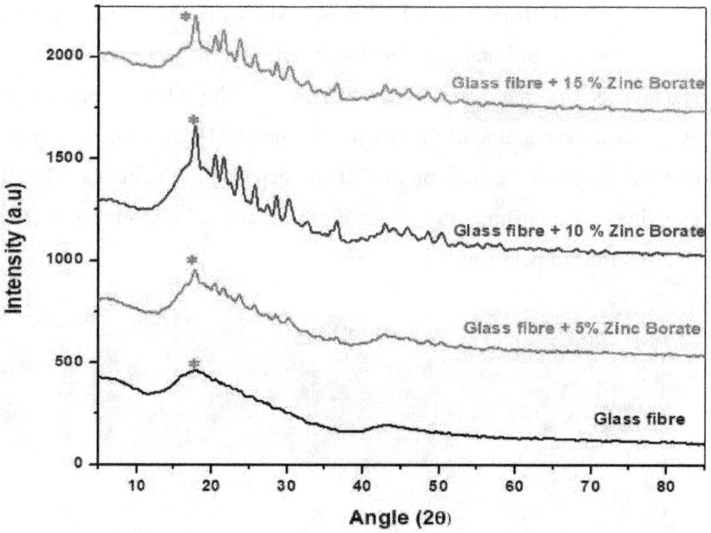

Figure 4.1 X-Ray Diffraction of Glass Fiber and Glass Fiber/Zinc Borate composite

In the XRD pattern, neither the raw GFRP nor the GFRP nanoparticles samples (0 and 5%) revealed any significant diffraction peaks. As a consequence of this, the absence of peak refraction in the nanocomposites was regarded as proof that the ZBP particles inside the epoxy matrix were successfully exfoliated. In the case of the 10 and 15 weight percent addition of zinc borate, there were distinct peaks that emerged at 17.8 of the 2-theta values and d-spacing values of 5 Å. When adding a greater proportion of nanoparticles, it became difficult to spread the particles due to their large propensity to agglomerate, which increases the viscosity of the resin.

Moreover, this tendency prevented the formation of the exfoliated structure. The peak intensity of X-rays increased after the incorporation of zinc borate, indicating the presence of cured epoxy chains. The 2-theta, which ranged from 10 degrees to 30 degrees, revealed the amorphous features of epoxy chains, attributable to their dispersion. The severity of the peaks reached its maximum at 10 weight percent. The addition of zinc borate established the bonding of fibers with zinc borate. Using a SEM, micrographs of glass fiber and the addition of ZBP are depicted in Figure 4.2. The EDAX experiment demonstrates the presence of zinc borate in the matrix of the glass fiber-reinforced polymer.

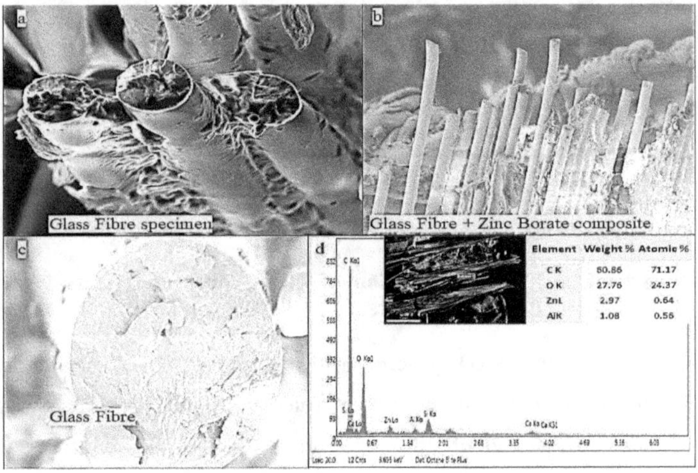

Figure 4.2 FESEM and EDAX of Glass Fiber and Glass Fiber/Zinc Borate composite

The SEM photos (Figure 4.2) provide evidence of a homogeneous dispersion since they show a uniform distribution of zinc borate particles throughout the GFRP matrix. Additionally, the EDAX analysis confirms that there are consistent levels of zinc and boron across many locations. Regarding the overall characteristics of the composite, such as its flame retardancy, this feature is highly beneficial. An enhanced comprehension of the interaction between the GFRP fibers and the matrix may be achieved through the analysis of SEM images.

Excellent adhesion is shown by a well-bonded surface that is free from gaps or debonding, as adhesion is crucial for weight transfer and overall strength. EDAX may be used to quantify the elemental ratios of zinc to boron, which is crucial for understanding the stoichiometry of the zinc borate phase. If the ratio deviates significantly from the stoichiometric ratio of zinc borate that is expected, it suggests that the reaction during the production of the composite was not fully completed. The SEM images reveal the presence of clusters or agglomerations of zinc borate particles inside the GFRP matrix.

The corresponding EDAX pictures exhibit areas with higher levels of zinc and boron, suggesting potential issues with dispersion as shown in Figures 4.3 (a) – 4.3 (d). The formation of these agglomerations may potentially diminish the mechanical properties and hinder the effectiveness of the flame retardant.

Furthermore, the utilisation of SEM in conjunction with EDAX can determine the presence or absence of a distinct zinc borate phase inside the composite sample. If a link exists between certain morphologies shown in the SEM images and regions containing significant amounts of zinc and boron, it suggests the occurrence of a distinct phase formation.

An analysis of the GFRP/Zinc Borate composite using SEM and EDAX provides crucial insights into the composite's microstructure, the distribution of additives, the bonding between different components, and the presence of any distinct phases. In order to obtain maximum performance, it is necessary to have a deep understanding of the development and production process of the composite material. This understanding is crucial for optimizing its qualities, including mechanical strength and flame retardancy.

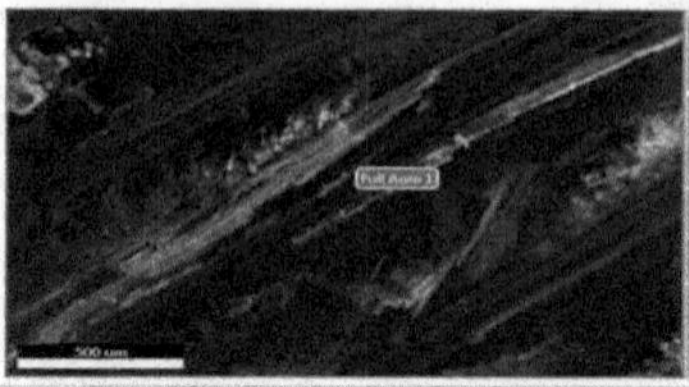

Element	Weight %	Atomic %	K ratio
C K	62.98	73.00	0.3952
O K	24.72	21.51	0.0853
MgK	0.48	0.27	0.0036
AlK	1.63	0.84	0.0128
SiK	6.33	3.14	0.0525
CaK	3.03	1.05	0.0251

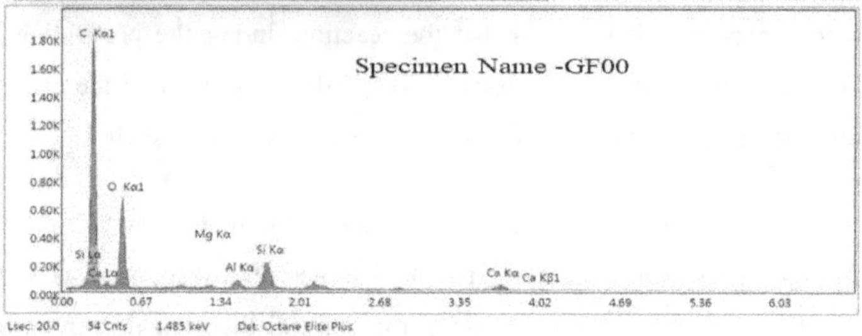

Figure 4.3 (a) EDAX of Glass Fiber without Zinc Borate composite 0 wt%

Element	Weight %	Atomic %	Kratio
C K	48.42	64.41	0.2630
O K	25.17	25.13	0.0958
ZnL	1.13	0.28	0.0078
MgK	0.86	0.56	0.0066
AlK	2.40	1.42	0.0193
SiK	9.08	5.16	0.0777
AuM	6.69	0.54	0.0428
CaK	6.26	2.49	0.0526

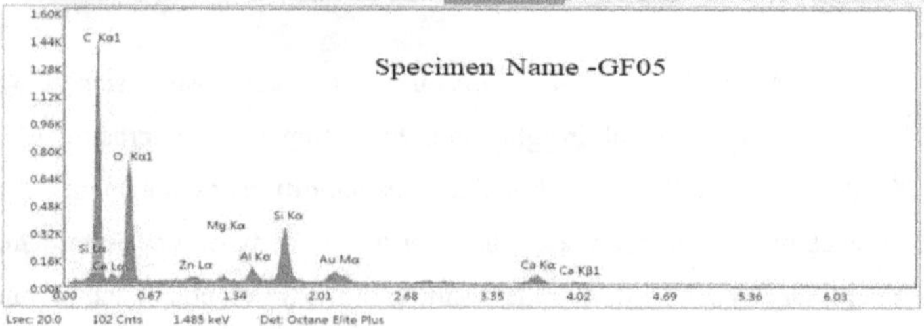

Figure 4.3 (b) EDAX of Glass Fiber and Glass Fiber/Zinc Borate composite 5 wt%

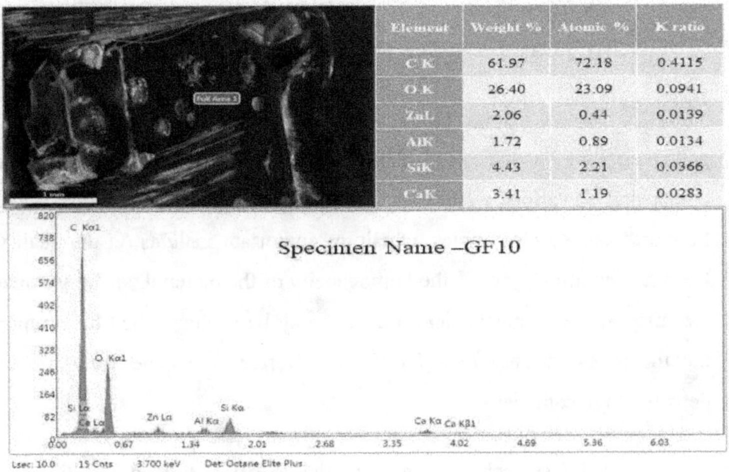

Element	Weight %	Atomic %	K ratio
C K	61.97	72.18	0.4115
O K	26.40	23.09	0.0941
ZnL	2.06	0.44	0.0139
AlK	1.72	0.89	0.0134
SiK	4.43	2.21	0.0366
CaK	3.41	1.19	0.0283

Figure 4.3 (c) EDAX of Glass Fiber and Glass Fiber/Zinc Borate composite 10 wt%

Element	Weight %	Atomic %	Kratio
C K	60.86	71.17	0.3962
O K	27.76	24.37	0.0996
ZnL	2.97	0.64	0.0199
AlK	1.08	0.56	0.0084
SiK	4.65	2.32	0.0383
CaK	2.67	0.94	0.0222

Figure 4.3 (d) EDAX of Glass Fiber and Glass Fiber/Zinc Borate composite 15 wt%

4.3 ESTIMATION OF THEORETICAL DENSITY AND ACTUAL DENSITY OF THE COMPOSITES

Table 4.1 presents the expected and actual densities, as well as the percentages of vacancies, of glass fiber and ZBP-filled glass fiber epoxy composites at various weight fractions. Additionally, the chart also includes the percentages of vacancies. Obtaining important insights on the quality of the manufacturing process, the homogeneity of the material, and any concerns that may affect its mechanical qualities may be accomplished by comparing the theoretical density of a GFRP/Zinc Borate composite with the actual density of the composite.

This research has the potential to offer direction for improving manufacturing operations in order to ensure that the composite material fulfils the design parameters that have been set within the project. The density of composites that was seen was lower than the density that was expected for them, which provided proof that voids were present inside the composites during the experiment. The density of GFRP that was measured was 1.55 g/cm^3, which was in contrast to the density that was anticipated, which was 1.63 g/cm^3. This is what Figure 4.4 illustrates. The densities of the composite are increased as a result of the incorporation of filler particles into the matrix. This is due to the fact that the filler particles have a higher density than the matrix. There was a direct correlation between the weight fraction and the density of the composites, which increased in tandem with the weight fraction.

The addition of zinc borate at a weight percentage of 10% led to a reduction in the percentage of voids, whereas the addition of zinc borate at a weight percentage of 15% led to an increase in the percentage of voids, as shown in figure 4.5. The findings of Sharma *et al*, (2020) reported were comparable. The proportion of voids in the GFRP construction was 4.91

percent. The formation of agglomerations occurred during the process of mixing and debonding the filler and matrix material. This led to a greater proportion of voids in the composites in comparison to the epoxy matrix. The filler quantity continued to be observed until it reached 15 weight percent.

Table 4.1 Theoretical density, actual density and void percentage of Glass Fiber / Zinc Borate reinforced epoxy composite

Fiber %	Resin %	Filler %	Theoretical density (g/cm^3)	Actual density (g/cm^3)	Void volume fraction %
50	50	0	1.63	1.55	4.91
50	45	5	1.674	1.60	4.42
50	40	10	1.716	1.66	3.26
50	35	15	1.756	1.65	6.03

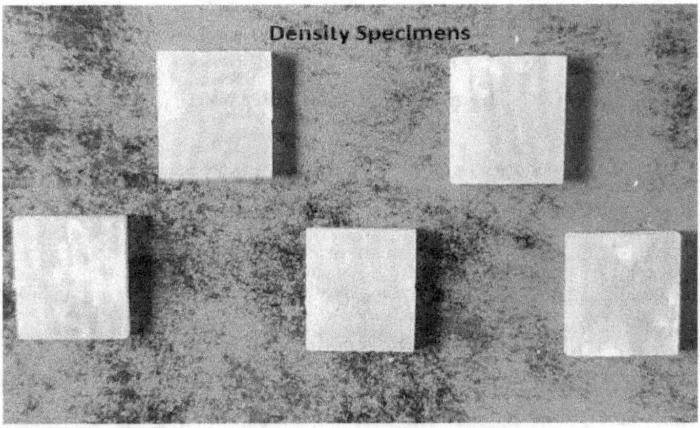

Figure 4.4 Specimens for Density test

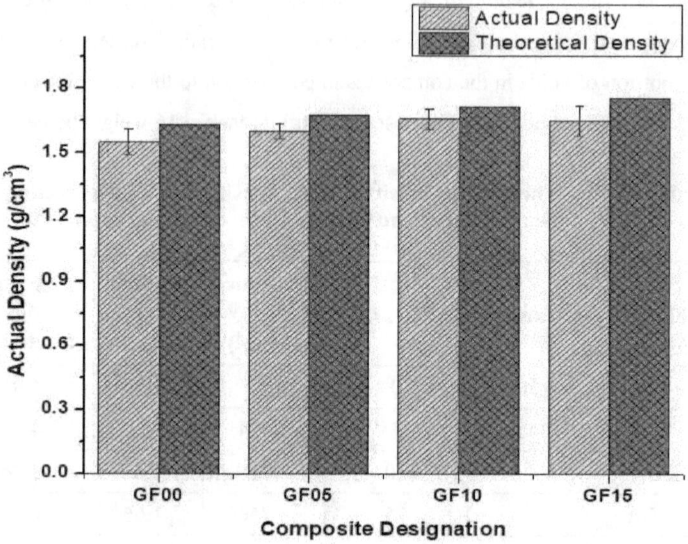

Figure 4.5 Theoretical density vs actual density of Glass Fiber/ Zinc Borate reinforced epoxy composite

4.4 SHORE D HARDNESS OF THE COMPOSITES

The Shore D hardness technique was utilized in order to determine the surface area of the sample that was resistant to being indented. Figure 4.6 displays the Shore D hardness metrics for the samples that were created according to the specifications. A direct proportional relationship existed between the degree of stiffness of each specimen and the Shore D hardness of the specimen. When compared to the GF00 composite, the sample that contained 10 weight percent zinc borate exhibited a 5.5% increase in its hardness. It was found that changes in zinc borate addition resulted in less significant disparities in hardness values. The results of the impact of ZBP on the resistance of composites to indentation are displayed in the outer layer of the glass fiber. When zinc borate particles were not added, the Shore D hardness was the lowest. This was followed by GF15, which had the highest addition. The agglomeration of zinc borate particles and the interaction between the matrix and reinforcement are the two factors that are responsible for the decrease in the degree of hardness.

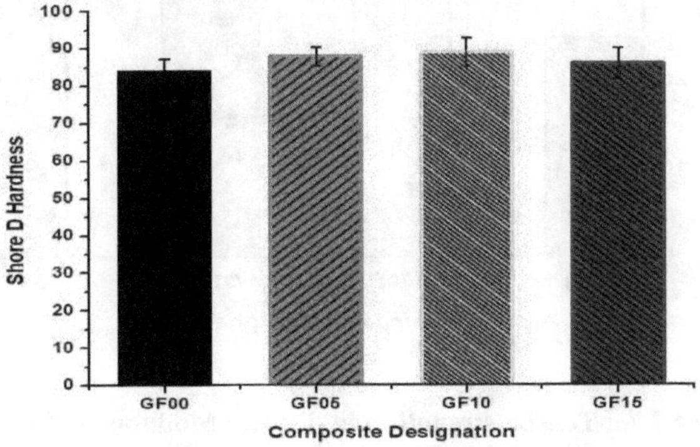

Figure 4.6 Shore D Hardness of Glass Fiber and Glass Fiber/Zinc Borate composite

4.5 ULTIMATE TENSILE STRENGTH OF THE COMPOSITES

The placement of the fibers and reinforcement impacted the ultimate tensile strength of the composites. The ultimate tensile strength values of the manufactured composites are illustrated in Figure 4.7. Fractured tensile test specimens with 5 wt. %, 10 wt. % and 15 wt. %. are illustrated in Figure 4.8. The GF10 composite exhibited the greatest ultimate tensile strength compared to other composites. The incorporation of ZBP at a concentration of 10 wt% demonstrated enhanced bonding between the matrix and fibers during fabrication, resulting in improved ultimate tensile strength. When a tensile load was applied to the manufactured composites, the outermost layer of fibers underwent a greater stress, which was subsequently transmitted to the underlying layers.

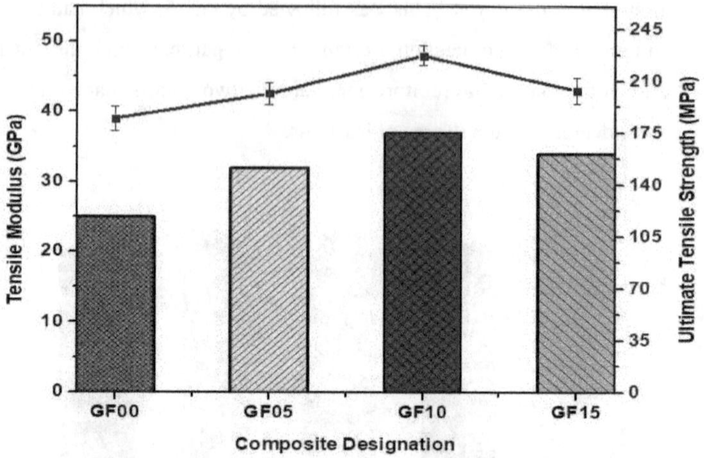

Figure 4.7 Tensile strength and Tensile Modulus of Glass Fiber and Glass Fiber/Zinc Borate composite

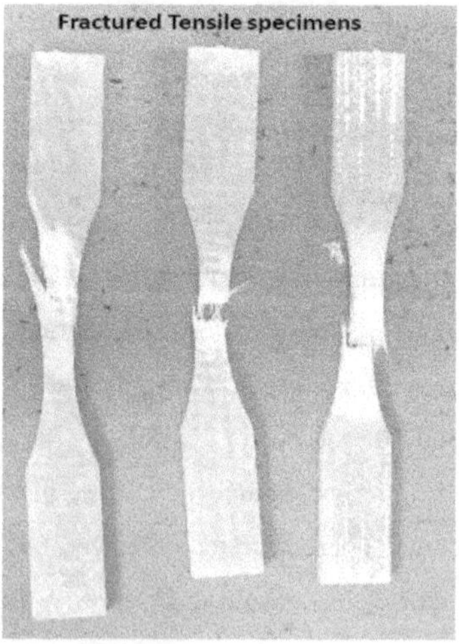

Figure 4.8 Fractured tensile test Specimens

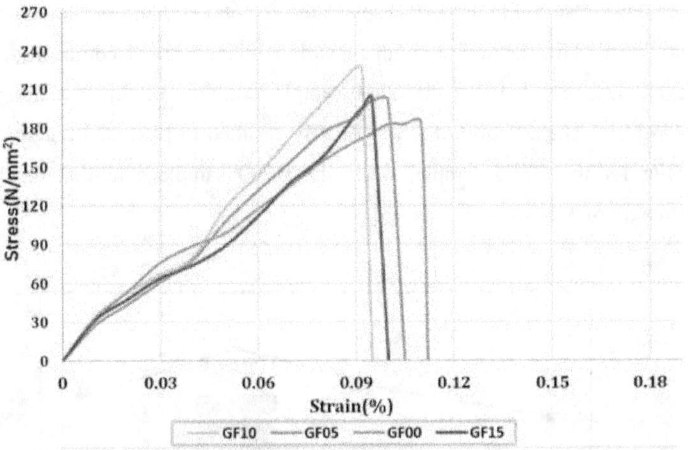

Figure 4.9 Stress-Strain graph of Glass Fiber and Glass Fiber/Zinc Borate composite

Zinc borate formed a bond between the resin and the matrix and provided interlocking strength. The lowest tensile strength was 185 MPa for the matrix and resin composite, revealing an improvement of 22%. There was a 10% decrease in tensile strength as the zinc borate reinforcement was increased to 15 wt. %. When zinc borate was added at 15 wt. %, it did not bind well with the glass fiber matrix, resulting in more voids and agglomeration of particles. The stress-strain graph shown in Figure 4.9 shows that higher strength was attained by a 10% addition of zinc borate. The fractured surfaces were more brittle in nature, with the fibers showing more resistance to fracture.

4.6 ULTIMATE FLEXURAL STRENGTH OF THE COMPOSITES

The stiffness of the composites was quantified using ultimate flexural strength. During the flexural test, the material experienced failure on the tension side. Figure 4.10 displays the flexural strength of the composites produced. The flexural strength of various manufactured composites, when evaluated, revealed that the GF10 sample exhibited the highest strength of 406 MPa. The addition of 5 wt. %. ZBP resulted in flexural strength of 375 MPa for the GF05 sample, while the GF00 sample exhibited the lowest strength of 353 MPa.

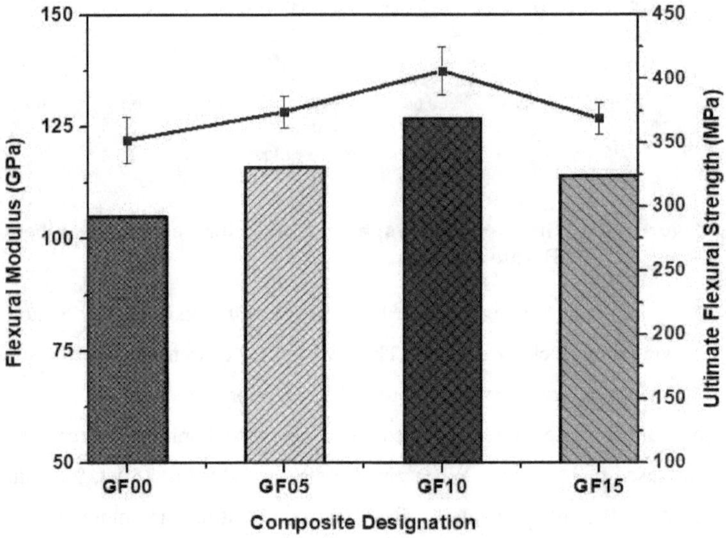

Figure 4.10 Ultimate Flexural Strength and Flexural Modulus of Glass Fiber and Glass Fiber/Zinc Borate composite

Enhancement of stiffness in the composites was observed when strong interfacial adhesion was established between the matrix and fiber. Nampoothiri *et al.* (2022) reported comparable enhancements in flexural strength due to the effective bonding between the matrix and the fiber. A decline in flexural strength was observed in other composites due to insufficient bonding interface between the matrix and the reinforcement. Prior research demonstrated that insufficient adhesion between the matrix and the fiber resulted in reduced flexural potency.

4.7 ULTIMATE COMPRESSIVE BEHAVIOR OF THE COMPOSITES

The ultimate compression strength of the composites manufactured is depicted in Figure 4.11. The compression test differs from the tensile test in that stress was applied in a vertical direction to the specimen. The application of stress vertically resulted in enhanced stress transfer, which consequently led to an augmentation in compressive strength. The composite material GF10, containing 10 wt. %. Zinc borate, exhibited the greatest compressive strength when compared to the other composites produced.

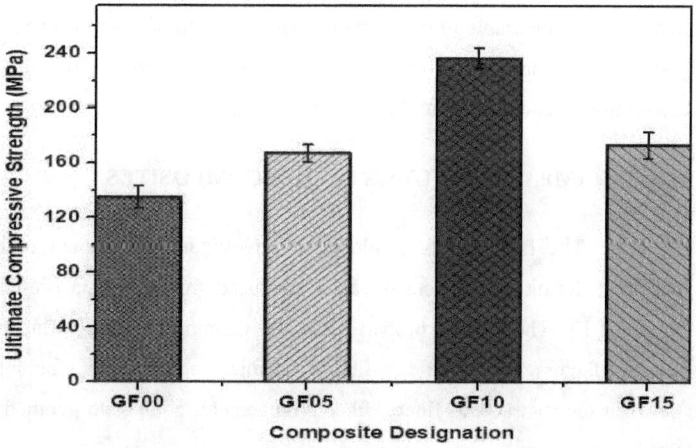

Figure 4.11 Ultimate Compressive Strength of Glass Fiber and Glass Fiber/Zinc Borate composite

Conversely, the composite lacking ZBP demonstrated the lowest compressive strength. The aforementioned outcome was evidence that the inclusion of ZBP had the capacity to absorb a greater quantity of compressive force. The enhanced transmission of stress from the fiber to the matrix was the primary rationale behind this phenomenon. The present empirical data was consistent with the outcomes documented by John *et al.* (2004).

Better interlocking improved compression strength with a strong link between the matrix and zinc borate nanogel reinforcement. Application of compressive load onto the matrix and resin composites resulted in a comparatively accelerated failure rate compared to other composites that incorporated ZBP.

The observed phenomenon was attributed to the delamination of Fibers, leading to a more rapid material failure and consequent reduction in compressive strength. Inadequate bonding between the matrix and reinforcement of the material lead to a reduction in its compressive strength, which was comparable to its ultimate tensile strength properties. The findings indicated that incorporation of ZBP leads to enhanced compressive force absorption in comparison to the GF00 composites.

4.8 IMPACT ANALYSIS OF THE COMPOSITES

The Izod test is employed to investigate the amount of energy absorbed during an impact, as can be deduced from the data presented in Figure 4.12. The impact properties of the composites were influenced by various factors, including the fiber's chemical composition, the interface between the matrix and fibers, fiber arrangement, composite geometry, and testing conditions.

In the current scenario, an abrupt load was imposed, and energy dissipation properties were determined. The GF05 composites exhibited the highest impact properties, followed by the GF10, as evidenced by the recorded data. The rise in the Izod test results could be attributed to the phenomenon of fiber pull-out. When identical fibers were stacked, there was a homogeneous distribution of impacts across all layers. The significant enhancement in impact resistance could be attributed to the possible dissipation of energy due to vibration and the propagation of fractures. The impact toughness of composite GF00 was observed to have a significant increase, which could be attributed to the presence of moderate fiber-matrix bonding.

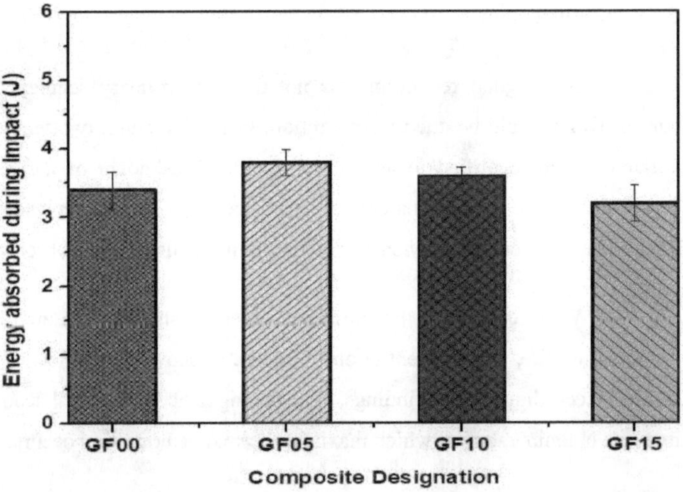

Figure 4.12 Energy absorbed during the Impact of Glass Fiber and Glass Fiber/Zinc Borate composite

This bonding is crucial to enable the material to absorb a greater amount of impact energy, surpassing the effects of incorporating 15 wt. % zinc borate. The findings pertaining to impact strength were consistent with the proposed results of Raghunathan *et al.* (2022). The incorporation of 5 wt. %. Zinc borate resulted in an 8.5% increase in impact absorption as compared to GF00.

4.9 THERMOGRAVIMETRIC ANALYSIS STUDIES

Figures 4.13 (a) and Figure 4.13 (b) illustrate both the derivative thermogravimetric (DTG) curves as well as the conversion rate vs. temperature for GFRP. Zinc Borate had a high percentage of thermal stability up to 307.3°C, 308.1°C, and 315°C, respectively, making it the highest of the remaining percentages. The findings, as illustrated in Figure 4.13 (a), were exhibited based on the alteration in weight of the specimens in response to the escalated temperature.

A marginal reduction was noted in the primary segment of the curves, which could be due to evaporation. In the literature, two temperature variables were identified to assess the thermal breakdown of the polymer composites. The initial parameter, denoted as T10%, represents the temperature that occurred when a reduction in mass of 10% was observed.

When enhancing the specimen's thermal stability, the zinc borate-based composite, at a concentration of 15 wt.%, showed significantly better results. According to the findings, introducing zinc borate will lead to an increase in temperature at which maximal thermal deterioration occurs.

There was a possibility that the improved thermal stability of composites could be attributed to an increase in the matrix material's crystallinity as well as the degree of cross-linkage present in the material. The relationship was between the amount of zinc borate and the temperature at which the compound decomposes. When there was a higher proportion of zinc borate present, the thermal decomposition temperature rose also.

However, this trend was reversed when there was a higher loading of ZBP; this was most likely due to the creation of compact char, which made it easier for composites when subjected to heat degradation. Decomposition started only when the substance reached a specific temperature. The composite containing 15 wt.% ZBP was, in comparison, the material with the highest thermal stability.

The DTG curve displays the rate at which the composite material's weight changes with temperature, offering insights into disintegration, deterioration, and phase transitions. The DTG curve of the GFRP/ZBP composite exhibits clear temperature transitions that provide insight into the characteristics of its constituent materials. The little decrease in weight observed at lower temperatures can be explained by the evaporation of any remaining moisture inside the composite. Subsequently, a prominent peak is detected in the graph representing the heat breakdown of the polymer matrix.

The presence of a distinct peak indicates a fast deterioration process of the polymer, most likely caused by the fragmentation of polymer chains and the liberation of volatile substances. Additionally, there is another prominent peak observed in the middle section of the curve, which suggests the breakdown of zinc borate additions in the composite material. The thermal event mentioned here is of great significance due to the common addition of zinc borate for its flame-retardant qualities.

Understanding the breakdown temperature of zinc borate is critical for comprehending how the composite material behaves when exposed to heat. The precise geographical coordinates and magnitude of the peak might offer valuable information on the efficacy of zinc borate in inhibiting the propagation of flames.

Moreover, an incremental decrease in weight seen at elevated temperatures correlates to the deterioration of the glass fibers. This phase is distinguished by a wide, steady incline in the DTG curve, indicating a slower and more progressive breakdown process in comparison to the polymer matrix and zinc borate. The inclusion of glass fibers has a substantial impact on the thermal stability of the composite, enhancing its structural integrity when exposed to high temperatures.

The DTG curve of the GFRP/Zinc Borate composite provides a comprehensive view of the thermal characteristics of its components. The curve clearly shows discrete peaks that correspond to the disintegration of the polymer matrix and zinc borate additives, as well as the slow deterioration of glass fibers. This research facilitates comprehension of the composite's thermal stability, flame retardant characteristics, and appropriateness for applications that include exposure to fluctuating temperatures.

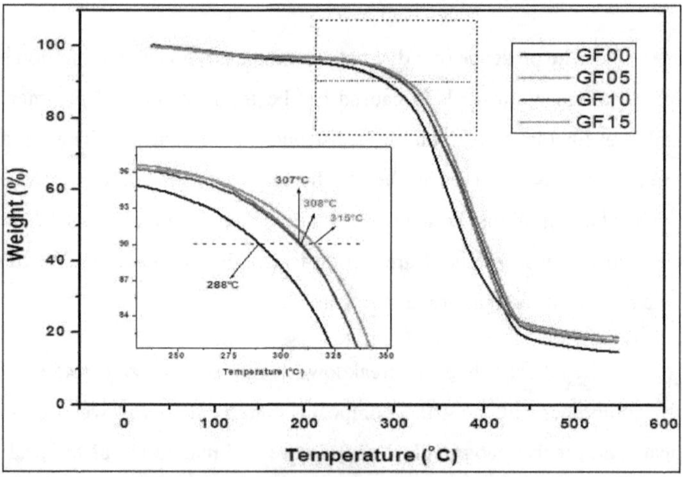

Figure 4.13 (a) Thermo gravimetric analysis of Glass Fiber and Glass Fiber/Zinc Borate composite

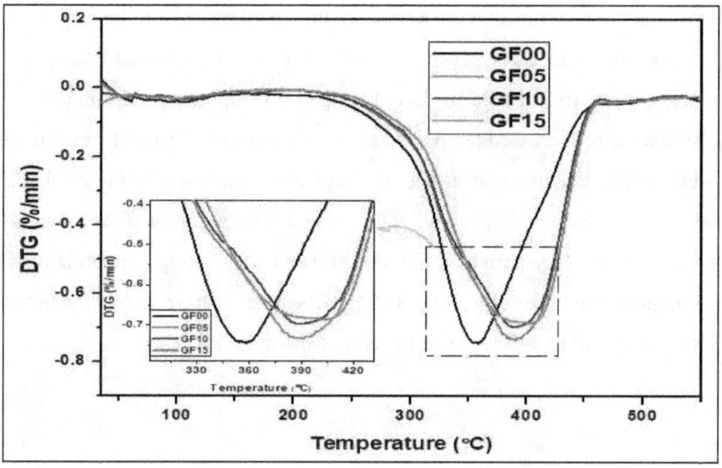

Figure 4.13 (b) Derivative Thermogravimetric curve of Glass Fiber and Glass Fiber/Zinc Borate composite

4.10 DIFFERENTIAL SCANNING CALORIMETRY STUDY

Differential scanning calorimetry, often known as DSC, is an effective approach that is utilized for the objective of analyzing the thermal characteristics of materials. This technique is utilized for composite materials, such as GFRP, which are coupled with compounds like zinc borate. The Tg, is an important factor in GFRP composites because of the influence it has on the mechanical characteristics of the composite material. The resin matrix is not considerably altered when GFRP composites are exposed to high temperatures below Tg, and the surface continues to be rough and comparable to the sample that has not been conditioned.

On the other hand, when the composites reach their Tg, the resin transforms from a glassy state to a rubbery one, which results in a significant decrease in both the strength and the stiffness of the composite substance. From the graphs of the various compositions, it can be seen that the addition of zinc borate resulted in a rise in the temperature of the Tg. In GFRP composites, the glass transition temperature is an important component that plays a significant role in determining the behaviour of the material when subjected to increased temperatures. The addition of 15wt. % zinc borate had influenced the increase in Tg of about 2%, as shown in Figure 4.14. This behaviour has an impact on the material's strength, stiffness, and overall mechanical characteristics. For the purpose of creating and utilizing GFRP composites in a variety of applications, it is vital to have a solid understanding of the Tg and its influence on the composite material.

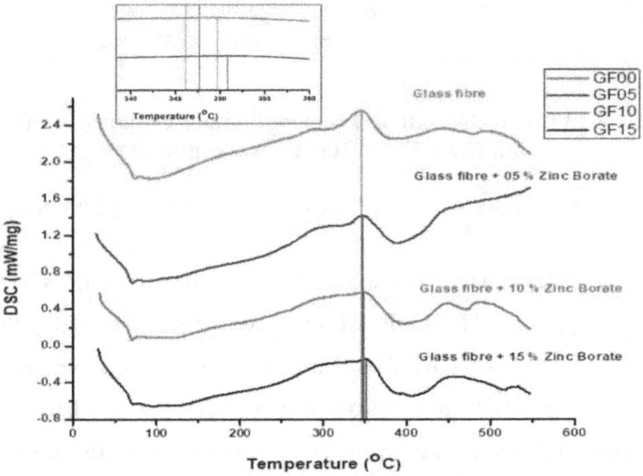

Figure 4.14 Differential scanning calorimetry vs Temperature

4.11 DYNAMIC MECHANICAL ANALYSIS

The mechanical and thermal properties are crucial for withstanding mechanical forces and ensuring safe handling in various practical scenarios. The DMA approach can be utilized to ascertain the thermo-mechanical characteristics of polymeric composites. The stated technique illustrates alterations in the mechanical properties of the material due to a suitable range of temperatures. When considering DMA, it is important to take into account the dynamic storage modulus (E'), loss modulus (E"), and tan δ (E"/E'). E' is associated with the elastic elasticity (Young's modulus) of the polymer, while E" is associated with the energy absorbed due to molecular motion inside the polymer. Determined the ratio of the loss modulus to the storage modulus to ascertain the contributions of the viscous and elastic components of a viscoelastic material. The secondary bonds of a polymer are constantly disrupted and re-established due to thermal motion.

Due to the influence of pressure, specific conformations are favored over others, causing the polymer molecules to gradually rearrange themselves into the preferred conformations. Thermal movement is a factor that affects the viscoelastic properties of polymers, causing changes in deformation characteristics when the temperature increases or decreases. Typically, when the temperature increases, the creep module is defined as the ratio of strain to stress, which is dependent on the stress. Typically, a rise in temperature decreases the duration required for the same level of stress to be applied under constant tension. Simply put, it requires less effort to stretch a viscoelastic material at a higher temperature compared to stretching it at a lower temperature.

The storage modulus of glass fiber-reinforced polymer composites plays a pivotal role in determining their mechanical properties and overall efficacy. Various experiments have shown that the storage modulus of these materials is influenced by multiple characteristics, such as fiber orientation, fiber length, and fiber-matrix adhesion.

The inclusion of glass fibers into the polymer matrix results in an increase in the storage modulus. This occurs because the inflexible glass fibers effectively transfer stress to the polymer chains, hence enhancing the overall stiffness of the composite material. The orientation of the fibers affects the storage modulus; aligned fibers typically produce larger values of the storage modulus than randomly orientated fibers. The storage modulus (E') quantifies the amount of energy stored in the elastic structure of a material, which is directly related to the material's elastic behavior.

This parameter is very advantageous for assessing the rigidity and elastic properties of composites. The relationship between the storage modulus and temperature for the produced composites is depicted in Figure 4.15. The storage modulus is also affected by the length of the fiber.

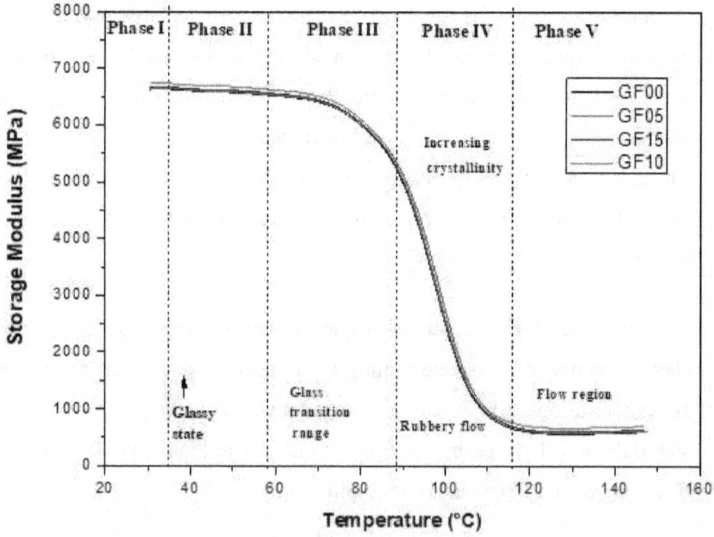

Figure 4.15 Storage modulus vs temperature

Greater fiber length enhances the ability to transfer stress to the polymer matrix, leading to higher storage modulus values. During the process of melting, there is a chance that the fibers may break, causing a drop in the actual length of the fibers and perhaps leading to a fall in the storage modulus. The elevated storage modulus of GF10 suggests that the glass fiber is the main factor responsible for the improved elasticity of the hybrid composite.

The adhesion between the glass fibers and the polymer matrix is another critical factor influencing the storage modulus. Strong adherence between fibers and the matrix facilitates effective transmission of stress, resulting in higher levels of storage modulus. By implementing surface treatments or incorporating coupling agents, the adhesion between the fibers and matrix can be enhanced, leading to an enhanced storage modulus of the composite. The graph shows that the inclusion of 10 wt. % nano Zinc borate particles resulted in an enhanced storage modulus compared to GF00 FRP composites.

The storage modulus curves exhibited minimal variation, and the subsequent inclusion of 15 wt% ZBP resulted in a decrease in the storage modulus, indicating a decline in the elastic characteristic of the hybrid composite. The glass transition state of the GF10 has been enhanced compared to GF00 and other samples due to the superior mixing of ZBP with the epoxy and glass fiber. The decrease in GF15 is caused by the clustering of the ZBP within the epoxy matrix, resulting in diminished molecular cohesion.

The storage modulus of glass fiber-reinforced polymer composites is influenced by various elements, such as the fiber content, fiber orientation, fiber length, and fiber-matrix binding strength. Acquiring understanding and refining these attributes can lead to the development of high-performance composites with improved mechanical properties.

When the temperature exceeds 110 °C, the storage modulus values of all the composites decrease to a minimum and then remain constant, indicating that the matrix has transitioned into a rubbery condition. In contrast to the storage modulus, the loss modulus (E”) is useful for assessing the viscous nature of a polymeric material. The parameter is quantified by measuring the amount of energy that is converted into heat and lost throughout a load cycle. The graph illustrates the loss modulus plots of unsaturated polyester composites, both single and hybrid reinforced, throughout a temperature range of 0 to 150 °C.

Evidently, all of the composites displayed comparable tendencies in the loss modulus curve. The temperature-dependent behavior of the material can be described as a progressive rise in the low-temperature range (<70 °C), a linear increase in the transition zone (80–90 °C), and a sudden decrease in the loss modulus at higher temperatures. The peak of the loss modulus curve can be utilized to calculate the Tg. The GF00 exhibited a lower loss modulus peak, indicating that the molecular chains possess greater mobility in the absence of ZBP reinforcement. The elevated peak values indicate that both composites, particularly GF10, exhibit a favorable fiber-matrix bond strength in comparison to the other composites.

All composites contained a greater proportion of glass fiber compared to the other composites, suggesting that the addition of glass fiber reinforcement has the ability to enhance the viscous behavior. Gupta and Deep also reported an increase in the viscosity response of glass fiber reinforced composites when using a greater glass fiber loading in the sisal/woven glass fiber reinforced polyester composite. While the addition of ZBP at greater composition did not result in any increase in the viscous response, it did enhance the thermal stability of the composites. This was shown through a shift in the peak towards a greater temperature range.

The improvement in thermal stability can be attributed to the restriction of segmental motion in the polymer chains provided by the reinforcing fiber. The tan δ, also known as the damping factor, is a ratio that quantifies the impact resistance and the relative contributions of the elastic and viscous phases of a material. It is calculated by dividing the loss modulus by the storage modulus. A high tan δ number signifies that the material exhibits greater viscosity and possesses a significant capacity for dissipating energy.

Conversely, a low tan δ value is indicative of a highly elastic material that may store energy within its structure. The Figure 4.16 illustrates the tan δ plots of the glass fiber reinforced polymer composites that were manufactured, covering a temperature range of 0 to 150 °C. The maximum tan δ values, as well as the Tg obtained from the tan δ and loss modulus curves, are summarized.

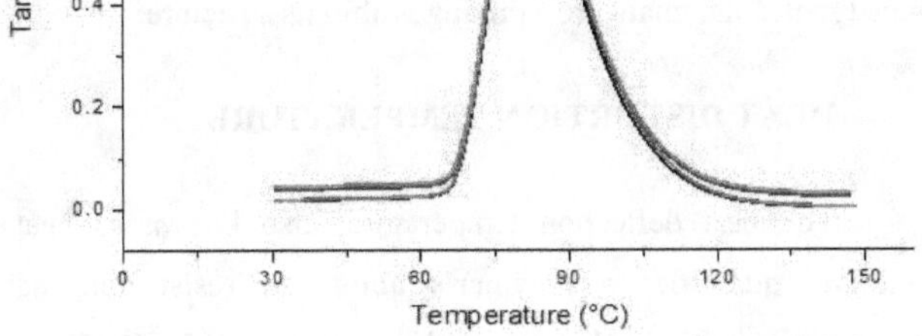

Figure 4.16 Tan Delta vs temperature

As demonstrated by the plot, GF10 exhibited the largest tan δ peak, which suggests a more viscous nature with increased polymeric chain mobility. The inclusion of both single and hybrid fiber reinforcements in the matrix has significantly enhanced the damping characteristic of the composites, resulting in reduced elasticity compared to the neat matrix GF00. This can be observed through the decreased peaks of tan δ. Out of all the reinforced composites, the GF00 composite exhibited the most substantial reduction in tan δ compared to the GF10 composite, with a 5% decrease. The GF15 composite exhibited a modestly reduced peak in tan δ, amounting to a 3% decrease compared to the GF10 composite matrix.

The inclusion of reinforcing fibers has resulted in limited mobility of the polymeric chains due to significant adhesion at the interface between the fibers and the matrix, thereby reducing the flexibility of the composite. Regarding the Tg obtained from the tan δ plot, it was seen that the addition of a single basalt fiber slightly enhanced the thermal stability, as indicated by the increase in Tg from 80.24°C to 81.37°C. Hybrid composites with greater loadings of ZBP in comparison to glass fiber resulted in higher Tg values. This discovery indicates that the addition of glass fiber leads to an enhancement in the thermal stability of the composites when combined with ZBP. The increased heat stability suggests that the composite possesses enhanced molecular chain crosslinking within its structure.

4.12 HEAT DISTORTION TEMPERATURE

The heat deflection temperature, also known as heat distortion temperature, quantifies a polymer's ability to resist deformation when subjected to a specific load at elevated temperatures. The HDT, as defined by ASTM D 648 (which is equivalent to ISO 75), represents the temperature at which a specific polymer test bar will deform by 0.25 mm under a specific load. The applied load value for the HDT is 0.46 MPa. During the test, the test bar is placed in a heated oil bath, resulting in a gradual increase in the temperature of the bath at a rate of approximately 2°C per minute. The deflection temperature, commonly referred to as the "deflection temperature under load" (DTUL), is a measure of a material's ability to withstand deformation under a certain load. This temperature is the maximum temperature at which a specific polymer may function without being damaged. Additional variables, such as the speed at which the temperature increases or the structure of the part, will significantly influence the overall thermal efficiency of a system.

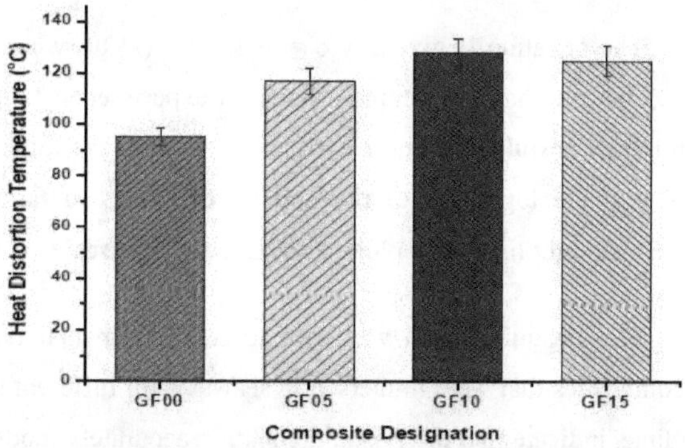

Figure 4.17 Heat distortion temperature for various composite designation

According to the graph, the HDT readings have improved as the reinforcement increased. The composite exhibits a high level of thermal stability, as evidenced by its resistance to bending caused by heat in oil. The highest HDT value of 128°C is observed when 10 weight percent of ZBP filler is added, while a value of 125°C is observed when 15 weight percent of ZBP filler is added, as illustrated in Figure 4.17. The increase in HDT values is directly correlated with the dispersion of fillers, whereas the decrease in temperature is attributed to the agglomeration of ZBP. Kojima *et al.* (2022) noticed a rise in HDT values when the percentage of fillers was increased.

4.13 FATIGUE STUDIES

The fatigue life prediction of randomly oriented long GFRP composites can be determined by applying the Basquin model, a widely used method for characterizing the S-N curve. The Basquin equation provides a mathematical relationship between the stress amplitude (σ) and the number of cycles to failure (N).

The equation is given by σ = A (2N) ^ (-1/b), where A and b are material constants that are obtained based on experimental fatigue data. The Basquin model postulates a linear correlation between the logarithm of stress (log (σ)) and the logarithm of the number of cycles to failure (log (N)), leading to a straight line when plotted on a logarithmic scale.

The Basquin model was used to fit experimental fatigue data for GFRP composites that were immersed in seawater for different time intervals. The findings indicate that the Basquin equation accurately models the fatigue characteristics in the high-cycle fatigue domain (N > 10^5 cycles), with the curve approaching an asymptote that can be regarded as the fatigue limit. Nevertheless, when considering low-cycle fatigue, it was shown that the Hwang and Han model provided a more accurate representation of the experimental data in comparison to the Basquin model. The Hwang and Han model is a description of the S-N curve that considers the non-linear correlation between stress and fatigue life.

Seawater exposure can have a substantial impact on the fatigue life of GFRP composites. The experimental findings demonstrated a direct correlation between the degradation rate, measured by the gradient of the normalized Wohler curves, and the duration of immersion in seawater. To summarize, the Basquin model is applicable for characterizing the S-N curve of GFRP composites in the high-cycle fatigue range, whereas the Hwang and Han model is more suited for the low-cycle fatigue range.

Nevertheless, the precision of these models might be affected by environmental variables like contact with seawater, which can result in deterioration of the composite material as time passes. The S-N curve for measured and model developed by Basquin equation (4.1) is represent in Figure 4.18.

$$\sigma_m = \alpha(N_f)^{\beta} \tag{4.1}$$

where 'σ_m' stands for mean stress, 'α' for coefficient of fatigue strength and 'β' for exponent of fatigue strength.

It can be seen from the graph that is displayed in Figure 4.18 that the fatigue strength grew significantly up to a nano filler addition of 10 weight percent, but then it decreased significantly as the nano filler content climbed further. The fatigue strength of the composite sample was the lowest when no filler particles were added. Reda *et al.* (2016) found that multi-walled carbon nanotubes can improve the fatigue life of GFRP by controlling the adherence of the matrix and preventing fatigue fracture development. Empirically, it has been proven that during fatigue cycling, cracks initiate and propagate in the matrix, leading to the occurrence of disbonds and delaminations. The continued spread of these fissures and separation of layers leads to catastrophic failure.

The initiation of disbonds and delamination in a nano-filler-added epoxy matrix can be anticipated by the progressive formation of fractures in the matrix. Moreover, a reduced rate of crack and delamination propagation effectively decelerates the overall failure process. Several micro-mechanisms, such as (i) suppressed matrix cracking, (ii) decreased crack propagation rates in the matrix, (iii) delayed the initiation of disbonds and delamination, and (iv) diminished delamination development rate, seem to extend the fatigue life of GFRP.

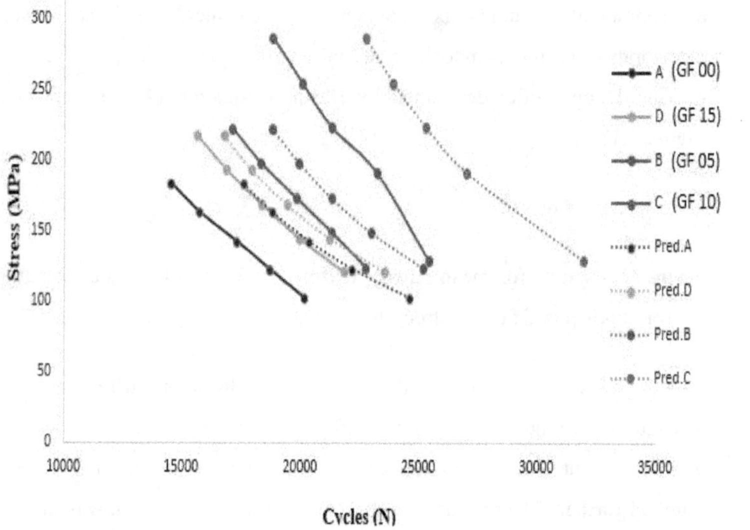

Figure 4.18 S-N curve of experimental and Basquin model

Also, the Figure 4.18 shows that the experimental findings and the fitted Basquin curve are in excellent accordance. The sample C (GF10) shows better results and the maximum number of cycles to failure. The sample D (GF15) which had the highest reinforcement of ZBP didn't show higher results due to the agglomeration of particles in the resin matrix causing voids and resin cracks. The emergence of transverse cracks in the weave's fill direction, followed by longitudinal inter-laminar cracks, indicates the beginning of fatigue failure during the first stage of fatigue life. Transverse matrix cracks typically originate from the specimen's edge, perhaps as a result of microdamage caused by the cutting process. As the repetitive stress persists, the number of cracks intensifies, leading to the subsequent occurrence of fracture coupling and separation at the interface.

Debonding can happen when the fibers that are oriented in a transverse direction separate from the matrix when breaks in the resin spread. The rapid decrease in modulus during this stage is mostly attributed to the propagation and buildup of cracks in several stress concentration sites.

4.14 WEIBULL FATIGUE LIFE MODEL

Many statistical processes and tests rely on certain distributional assumptions. The assumption of normality is highly prominent in traditional statistical tests. Dependability modelling often assumes that the data will conform to a Weibull distribution. The Weibull fatigue life model can be an effective method for forecasting the fatigue life of GFRP composites. In particular, this proved effective to analyze bending fatigue life of GFRP (RK, Sundaram *et al* (2022)). This approach offers valuable insights into the fatigue characteristics of the material when subjected to cyclic loading circumstances. The Weibull fatigue life model can be used to evaluate the bending and radial compression strength of composite rods reinforced with Zinc borate in the context of GFRP. The model enables the identification of fundamental fatigue properties and the creation of S-N curves, which are crucial for comprehending the fatigue behavior of GFRP composites reinforced with Zinc borate.

Moreover, the utilization of the Weibull fatigue life model can assist in examining the causes of failure under both static and cyclic loading conditions in bending tests. The model can offer useful insights into the material quality and behavior under fatigue circumstances by examining the microstructure of the rebars and detecting failure processes including delamination and fiber fracture.

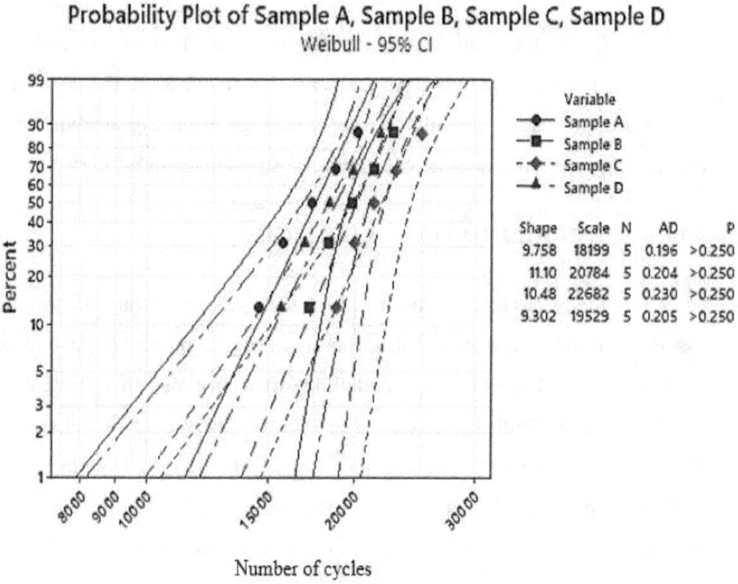

Figure 4.19 Weibull probability plot for samples A, B, C, D

The Weibull fatigue life model is a reliable analytical technique for evaluating the fatigue performance of GFRP composites with Zinc borate reinforcement. It provides a systematic method for comprehending the material's durability and fatigue behavior under different loading circumstances. The specimen GF00 (sample A) had shown the least no of cycles to failure as compared to the GFRP composites with ZBP reinforcement. The best result was shown with the 10 wt. % addition of ZBP particles and it's due to the improved interfacial bonding between the long glass fibers and the resins mixed with ZBP. Sample D (GF15) exhibited the most significant increase, demonstrating a reduction in the fatigue cycle attributable to brittleness and fractures that occurred in the resin cracks, as illustrated in Figure 4.19.

The resin fracture and the fiber pull out due to cyclic loading on the specimen is discussed in the fractography study of the specimens. The AD test, also known as the Anderson-Darling test, is a statistical method used to assess data related to fatigue. It calculates a P value, which provides information on the significance of the observed data. The magnitude of the Weibull probability distribution function is statistically determined by the Anderson-Darling test (A-D test). The value can be obtained by calculating the square of the distance between the straight line and the depicted point. The A-D statistical test is utilized to determine the optimal distribution for the collected data. Using MINITAB software, it is feasible to determine the A-D value by analyzing the goodness of fit distribution. This technique is used to extrapolate the results to situations that are representative of typical use.

The similarity of the regression lines under accelerated conditions allows for the determination of the acceleration between the two specified environments. There is a relationship between the form parameter of the Weibull distribution and the slope of the regression line. The way of verifying acceleration is by validating the similarity of the slope of the regression line, as previously said. The figure 8 displays the value of accelerated fatigue life. Consequently, the regression lines are all parallel, the Weibull distribution is established with a 95% confidence level, and it is believed that the acceleration of the fatigue life data has been accomplished.

4.15 FRACTOGRAPHY STUDY OF GLASS FIBER/ZINC BORATE SPECIMENS

Scanning electron microscope micrographs in Figure 4.20 show that the presence of strongly processed glass fiber fillers along the crack route significantly slowed the crack's progression during the tensile loading and fatigue tests. The epoxy matrix's increased stiffness and modulus may be traced back to the addition of zinc-borate fillers; this was accomplished by limiting the mobility of the polymer chains in the matrix. The morphological study of composite materials like zinc boron reinforced with glass fibers involves examining the arrangement, distribution, and interactions of the different components to understand their effects on the material's properties and performance.

Figure 4.20 Fractography studies of the fractured specimens of Glass Fiber and Glass Fiber/Zinc Borate composite

The agglomerated nano-ZBR particles act as stress concentrators and can reduce the mechanical properties. The fracture was effectively deflected due to the filler's apparent homogeneous distribution across the epoxy matrix. The surface of the fracture of filler-loaded epoxy seemed unevenly rough and coarse, while the crack course looked deformed and perturbed/tortuous, proving this point. In response, the crack deflection mechanisms shifted the stress condition by tilting or twisting the fracture to redirect it away from the initial propagation plane.

Further, the filler's presence bridged the fracture front, and the strong interfacial interface between the filler and matrix necessitated extra energy for the crack to propagate further, leading the filler to break through, potentially boosting the material's toughness. The presence of cohesive failure at the filler/matrix interfaces, where the matrix fragments remained attached to the filler particles, indicated a potential for increased energy dissipation before fracture through toughening mechanisms such as filler debonding or pullout. These mechanisms were directly linked to the plastic zones caused by matrix shear yielding, followed by the de-bonding process.

Scanning electron microscopy images indicated that the proportion of empty spaces in this 10-wt% addition was generally quite minimal. The presence of air engulfed encircling glass Fiber fillers may account for the slightly higher values observed in 15 wt. % ZBP filler specimens compared to those with 10 wt. % ZBP filler. The adhesive failure of the exposed fiber layers suggested insufficient adhesion between the fiber and the matrix.

Samples containing 5 wt. % weight Zinc Borate filler exhibited a fracture surface with a rough texture, indicative of ductile failure. In the case of 15 wt. %, there was a decrease in characteristics due to the pull-out of fibers. It happens when the bond between the fiber and the matrix material is inadequate to withstand the applied load. Due to more voids present in

15 wt. % addition, the cracks propagate and pull the fiber from the sockets inside the matrix, as it is the primary mechanism responsible for enhancing the toughness of fiber composites.

The inclusion of filler material impeded crack advancement, thereby inhibiting the amplification of micro cracks within the matrix through the formation of extensive hackles. This mechanism ultimately improved the material's interlaminar fracture. Typically, the enhancement of fracture toughness was attributed to toughening mechanisms such as debonding, crack deflection, and filler/matrix interface interlocking, which need a greater amount of energy to initiate or promote a crack.

4.16 SUMMARY

In order to ensure the presence of zinc borate in the fabricated composite specimen an in-depth discussion on EDAX was presented in this chapter, as well as the various mechanical characterization results, thermal characterization results, and fatigue study results for the composite specimens that were manufactured using the compression moulding process, were all described. The fractography of the composite specimens taken with SEM images is also explored in detail.

CHAPTER 5

ABRASIVE WEAR STUDIES OF GFRP/ZBP COMPOSITE

5.1 INTRODUCTION

Polymers have begun to exploit their advantageous strength-to-weight ratio and corrosion resistance, making them a viable substitute for many engineering materials. These polymers are being used in many applications, such as biomedical, bearings, belts, seals, brakes, and more. This chapter provides a detailed analysis of the tribological properties of GFRP/ZBP composites that have been produced by manufacturing processes. The tribological characterization is conducted by an abrasive wear test. The tribological properties of GFRP/ZBP composites are assessed using 320-grit SiC abrasive paper. The findings from the abrasive wear test have been organized into a table and analyzed. The SEM is used to visually analyze the abraded surface in order to identify the primary wear mechanism observed during the abrasive wear test.

5.2 MATERIALS AND PROCESS PARAMETER SELECTION FOR ABRASIVE WEAR TEST

The GFRP/ZBP composites with weight percentages of 5, 10, and 15 were chosen for the abrasive wear test. Abrasive paper with a grit size of 320 was adhered to a revolving disc made of EN 8. During the abrasive wear test, the following criteria were deemed significant: (i) Percentage of ZBP filler (wt. %), (ii) Applied load (N), and (iii) Sliding distance (m). Figure 5.1

displays the SEM picture of paper with a grain size of 320. Figure 5.1 clearly displays the homogeneous dispersion of SiC particles.

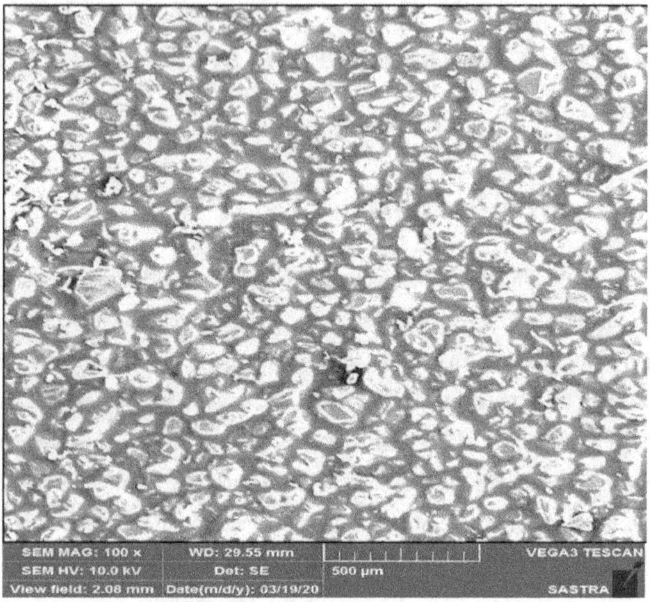

Figure 5.1 SEM image of 320 grit size paper

5.3 DESIGN OF ABRASIVE WEAR TEST EXPERIMENT BY TAGUCHI METHODOLOGY

Optimizing process parameters is a laborious effort, and as the number of parameters increases, it becomes progressively more difficult to carry out these enhancements. As the quantity of variables expands, there is a concomitant rise in the quantity of experiments. To address these limitations, the Taguchi technique utilizes orthogonal arrays as a design, resulting in a reduction in the number of required tests. Abrasive wear tests are conducted in a dry environment, utilizing factors such as the weight percentage of zinc borate filler, the applied force, and the sliding distance. These tests are performed at a constant sliding velocity of 0.5 m/s. Table 5.1 provides a comprehensive list of the control components used in the abrasive wear test, along with their corresponding values. Tests are conducted on an L16 orthogonal array to examine the relationship between input parameters (% of zinc borate filler, applied load, and sliding distance) and output parameters, including weight loss, coefficient of friction (CoF), and specific wear rate (SWR).

Table 5.1 Control factors and their levels of abrasive wear test

Control factors	Levels			
	1	2	3	4
Percentage of Zinc borate filler (wt. %)	0	5	10	15
Applied load (N)	05	10	15	20
Sliding distance (m)	125	250	375	500

5.4 STUDY ON ABRASIVE WEAR CHARACTERISTICS ON SPECIFIC WEAR RATE AND COEFFICIENT OF FRICTION

The abrasive wear test was carried out according to L_{16} orthogonal array and the values of CoF and SWR are listed in Table 5.2. All experiments were carried out at a constant sliding velocity of 0.5 m/s.

5.4.1 Weight Loss of Abraded Samples of GFRP/ZBP Composites

The proportion of ZBP nano-particulate (wt. %), the sliding distance (m), and the load that was placed (N) are all factors that are taken into consideration in Table 5.2 to determine the weight loss. In accordance with the findings, the weight reduction is largest when the percentage of ZBP-particulate in GFRP/ZBP composites is adjusted to zero weight percent. This is in comparison to the other weight percent of ZBP particulate in GFRP/ZBP composites. GFRP/ZBP composites that include a greater quantity of ZBP particles lead to a reduction in weight loss.

Table 5.2 Experimental findings of specific wear rate and coefficient of friction at 0.5 m/s sliding velocity

Sl. No	% Rein	Load (N)	SD (m)	Time (s)	Specific wear rate (mm^3/Nm)	COF
1	0	5	125	4.16	0.0061	0.27
2	0	10	250	8.33	0.0057	0.26
3	0	15	375	12.5	0.0025	0.243
4	0	20	500	16.59	0.0018	0.2205
5	5	5	250	8.33	0.0064	0.253
6	5	10	125	4.16	0.0054	0.245
7	5	15	500	16.59	0.0024	0.226
8	5	20	375	12.5	0.0019	0.2241
9	10	5	125	4.16	0.0050	0.2324
10	10	10	500	16.59	0.0024	0.2169
11	10	15	375	12.5	0.0023	0.22
12	10	20	250	8.33	0.0028	0.2214
13	15	5	500	16.59	0.0023	0.2187
14	15	10	375	12.5	0.0033	0.2124
15	15	15	250	8.33	0.0029	0.2169
16	15	20	125	4.16	0.0024	0.2

Given that they both grow with each other as seen in Figure 5.2(a). Sliding distance and applied force are two of the most important factors for weight reduction. There is a maximum weight loss of 0.0128 grams when the reinforcement of ZBP nano-particle in GFRP/ZBP is at a weight density of 5%. There is a maximum weight loss of 0.0091 grams when the reinforcement of ZBP particle in GFRP/ZBP is at a weight percentage of 15%.

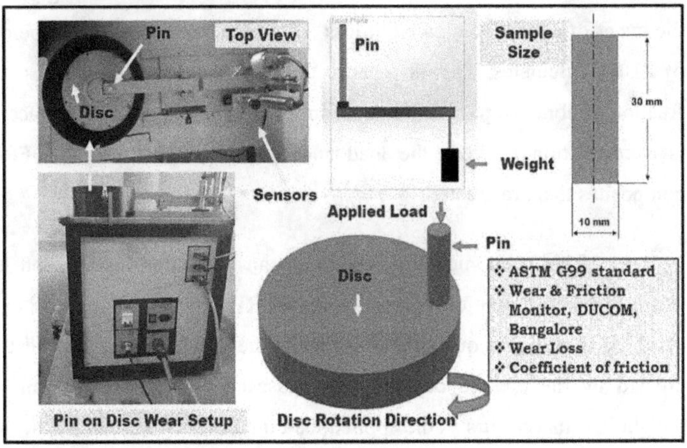

Figure 5.2(a) Schematic diagram of the abrasive wear process

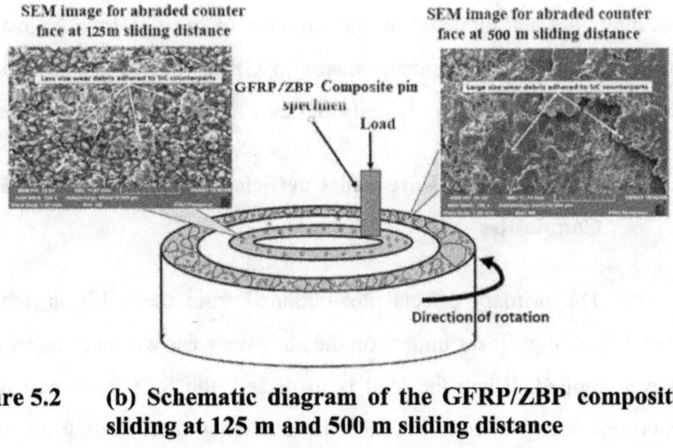

Figure 5.2 (b) Schematic diagram of the GFRP/ZBP composites pin sliding at 125 m and 500 m sliding distance

An improvement in wear resistance may be observed in GFRP/ZBP composites when the percentage of ZBP particulate is 15 weight percent. This is in contrast to other percentages of ZBP filler in GFRP/ZBP composites, which have a lower percentage of ZBP filler. The construction of an agglomerated structure, which serves as a barrier against deformation, is seen in Figure 5.2(b). This structure is created during the manufacturing process. In the event of a 10 N load, weight loss is lowered regardless of the percentage of ZBP particulates. This is because there is less penetration and a smaller fraction of abrasive particles that are in touch with the rubbing surface. There is a correlation between the load and the weight loss of the GFRP/ZBP composites that are created.

This phenomenon arises due to an expansion in the depth of SiC grit infiltration on the outer layer of the GFRP/ZBP nanocomposite material, resulting in a higher quantity of grooves. The greater amount of heat that is created by the pressed contact zone causes an increase in the amount of weight loss that occurs as the applied load increases. Additionally, the surface layer of GFRP/ZBP nanocomposite gets plastically deformed as a consequence of the heat that is created in the contact zone. This leads to debonding and an increase in the amount of weight that is lost. When comparing to other ZBP enhancements to GFRP matrix, the weight lost by 15% GFRP/ZBP composites reveals that they are more resistant to wear.

5.4.2 Specific Wear Rate and Coefficient of Friction of GFRP/ZBP Composites

The primary effects plot obtained from the SWR is depicted in Figure 5.3. A significant impact on the abrasive wear was exerted by the load that was applied. When the load is increased, the SiC paper will penetrate deeper into the polymers because the frictional force will grow in direct proportion to the load and the grit particles. Due to this, there is an increase in the creation of grooves as well as wear and tear on composites. Starting with 5 N and progressively increasing to 20 N, the fraction of the applied load that is attributed to wear loss is steadily raised. It is observed that the specific wear rate is greater at 10 N and 15 N when the loading condition is 20 N. If we compare the particular wear rates of 5 N and 10 N, we find that there is a difference of 55% in percentage. Similar to the previous example, the percentage difference between 10 N and 15 N is 30%, and the difference between 15 N and 20 N load is 17%. According to the sliding distance and grit size of SiC abrasive paper, the mean value of wear rate increases linearly during the application of load, and it is strongly dependent on both of these factors. Because of this, there is an increase in the creation of grooves and the wear and tear of composites in relation to the load that is applied.

It is seen that the particular wear rate graph tends to demonstrate a decrease in the percentage of wear loss that is caused by the buildup of wear particles between the SiC paper and the specimen. This, in turn, results in a reduction in the ongoing impact of SiC penetration and contact stress that is caused by the load that is given to the specimen.

5.4.3 Specific Wear Rate of Abraded GFRP/ZBP Composites

The main effect plot of SWR with respect to percentage of ZBP-filler (wt. %), sliding distance (m) and applied load (N) is shown in the Figure 5.3.

5.4.3.1 The influence of sliding distance on the specific wear rate of abraded GFRP/ZBP composites

The created GFRP/ZBP composite specimens demonstrate a high SWR up to 250 m during the first phases, and as the sliding distance increases, the values of SWR fall steadily. The GFRP/ZBP composites exhibit increased surface wear resistance (SWR) when the outer layer is exposed at short abrading distances, specifically ranging from 125 m to 500 m. The discrepancies in wear values seen during abrasion at different distances are attributed to the inability to obtain the whole wear track.

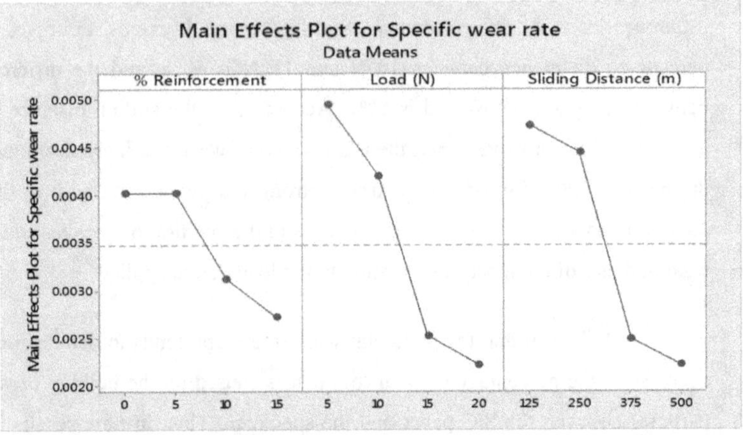

Figure 5.3 Main effects plot for specific wear rate (mm^3/Nm)

As demonstrated in Figure 5.2, when the abrading distance is very short, the clogging of abrasive wear debris in the 320-grit paper is very limited. When the pin is moved further away, it continually glides over the same track for a long time. This creates a film on the abrasive surface, which reduces friction and improves lubrication. As a result, the SWR of GFRP/ZBP composite specimens is lower.

5.4.3.2 Effect of applied load on specific wear rate of abraded GFRP/ZBP composites

As shown in Figure 5.3, as the applied load is increased, the SWR values drop, regardless of whether or not there is an increase in the proportion of ZBP nano-particulate reinforced. The GFRP/ZBP composites can easily separate from the surface when a higher applied load is exerted due to the rise in heat generated in the contact zone. This is due to the plastic deformation of the GFRP/ZBP composites surface caused by the increase in heat produced in the contact zone. The formation of a transfer layer on the 320-grit paper is the primary cause of the decrease in SWR values. Using a load of 15 N, this investigation found that SWR was reduced by half.

5.4.3.3 Effect of percentage of ZBP nano-particulates on specific wear rate of abraded GFRP/ZBP composites

The addition of ZBP nano particulates to the matrix of GFRP/ZBP composites results in a greater SWR at lower percentages. However, as shown in Figure 5.3, the SWR decreases as the fraction of ZBP particles increases. The chemical reaction that takes place between ZBP and GFRP results in increased adhesion between the two materials as a direct result of the contact that takes place between them. Because of the creation of aggregates in the GFRP/ZBP composites, the SWR values fall when the ZBP particle percentage approaches 15 wt.%. This is due to the accumulation of aggregates. Consequently, experiments have shown that adding five weight percent of ZBP to GFRP/ZBP composites leads to a decrease in the composites' shear strength.

5.4.4 Coefficient of Friction of Abraded GFRP/ZBP Composites

Figure 5.4 illustrates the CoF as a function of the variation in applied load. It demonstrates that as the applied load increases, the CoF decreases in magnitude. The observed phenomenon can be attributed to a disparity in temperature between the surface of the GFRP/ZBP composite specimen and the contact zone consisting of 320 grit. When a applied load of 20 N is applied, the friction between the surfaces causes heat to build up, leading to the surface of the GFRP/ZBP composite partially melting. This process results in the lubricating of the interface, which in turn decreases the

CoF values. The main effect plot of CoF with respect to percentage of ZBP particulates (wt. %), sliding distance (m) and applied load (N) is shown in the Figure 5.4.

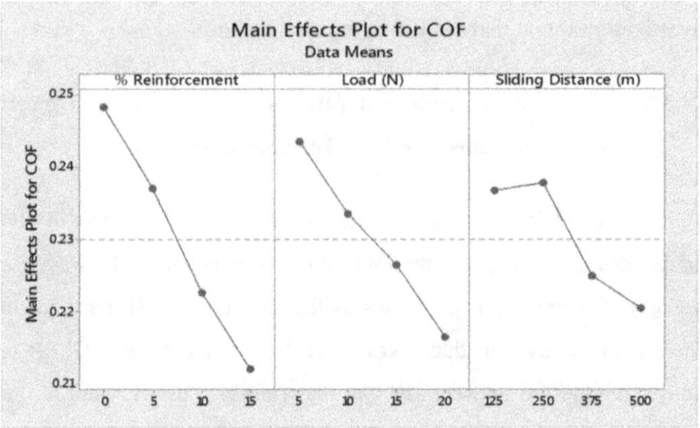

Figure 5.4 Main effects plot for coefficient of friction (μ)

5.4.4.1 Effect of percentage of ZBP particulates on coefficient of friction of abraded GFRP/ZBP composites

Figure 5.4 illustrates the impact of CoF on the proportion of ZBP particles (wt. %) as a function of variance. Friction arises prior to attaining a state of equilibrium due to changes in the contact area and shear strength of the polymer. The majority of the applied load is distributed between the ZBP and the GFRP matrices. The CoF is greatest in the scenario involving 5 wt. % GFRP/ZBP composites, and it lowers as the amount of ZBP particles increases, as seen in Figure 5.4. The decline in the CoF can be attributed to the enhanced thermal stability seen in the 15 wt. % GFRP/ZBP composites, as determined by TGA testing.

5.4.4.2 Effect of applied load on coefficient of friction of abraded GFRP/ZBP composites

The fluctuation of the CoF with regard to the applied load is depicted in Figure 5.4, which shows that the value of the CoF decreases with an increase in the applied load. This phenomenon happens as a result of the temperature disparity between the surface of the GFRP/ZBP composite specimen and the contact zone with a 320-grit surface. At a force of 20 Newton, frictional heat builds up at the interface, resulting in partial melting of the GFRP/ZBP composites. This melting acts as a lubricant, reducing the CoF.

5.4.4.3 Effect of sliding distance on coefficient of friction of abraded GFRP/ZBP composites

Figure 5.4 depicts the correlation between the extent of sliding distance and the value of the CoF. When the sliding distance is extended to 150 meters, there is a significant decrease in the CoF values, which is followed by a substantial drop in the CoF value beyond this sliding distance. The rate of abrasive wear decreases as the sliding distance increases due to the accumulation of debris between the cavities of 320 grit paper. This is due to the increased resistance encountered when moving the paper. The diminished efficacy of the material when using 320 grit ultimately leads to a decrease in the extent of abrasive deterioration. Furthermore, the occurrence of reduced effectiveness can also be described as abrasive wear clogging. This is an additional factor contributing to the blockage caused by abrasive wear.

Figure 5.5 displays the interaction plot depicting the relationship between % reinforcement, applied load, sliding distance, and specific wear rate. The data reveals that the wear resistance exhibits a higher magnitude when subjected to low loads and when the distance of trip over the abrasion surface is reduced.

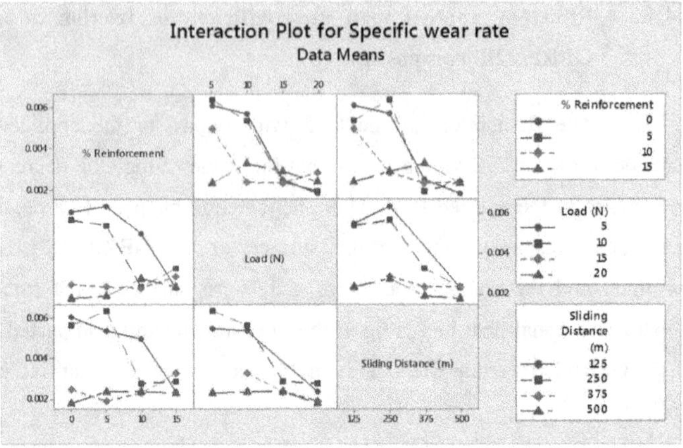

Figure 5.5 Interaction plot for ZBP vs Applied Load vs Sliding Distance for Specific wear rate

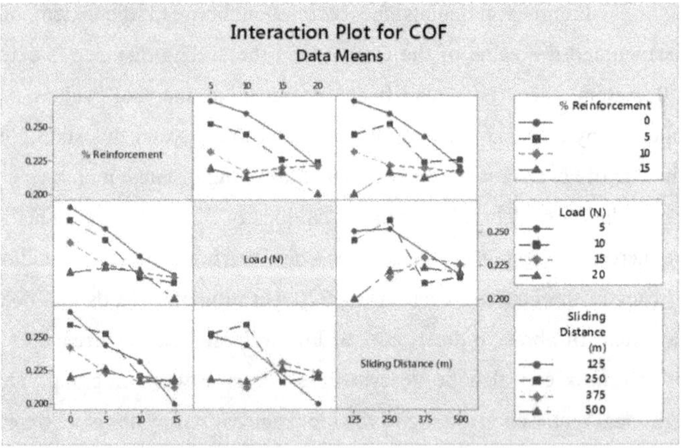

Figure 5.6 Interaction plot for % reinforcement vs Applied Load vs Sliding Distance for coefficient of friction (μ)

The wear rate escalates when exposed to increased loads that range from 15 to 20 N and sliding lengths varying from 375 to 500 m. With a load of 5 N and a sliding distance of 125 m, the specific wear rate decreased. No noticeable increase in wear was noticed in the applied load until it slid a distance of 250 m. The decreased wear is a result of the pin covering a shorter distance on the SiC paper. The wear resistance with high loads and sliding distance has been linked to the ZBP particles.

An interaction plot has been developed for CoF, illustrating the relationship between ZBP reinforcement, applied load, and sliding distance. According to the findings shown in Figure 5.6, it can be observed that the CoF is greater under conditions of low applied load, 5% filler content, and minimal abrasive sliding distance. One can analyze the correlation between the parameters that cause wear. The graph depicting the applied force and sliding distance provides strong evidence of the prevalence of wear in the context of abrasion. The reinforcement with a weight percentage of 0 (pure GFRP) exhibited more wear, whereas the addition of reinforcements resulted in a decrease in the CoF value, particularly at 10 wt. % and 15 wt. % reinforcement of ZBP.

5.5 ANALYSIS OF VARIANCE FOR SPECIFIC WEAR RATE AND COEFFICIENT OF FRICTION

The ANOVA approach is employed to assess the impact of input elements, such as the proportion of ZBP filler, sliding distance, and applied load, on the output findings (CoF and SWR) by analyzing the relationship between various variables in the data. The objective of the ANOVA is to identify the abrasive wear components that exert a statistically significant influence on the output. This study used the Minitab V 16.0 program to assess aspects such as lower-the-better quality, the CoF and the SWR.

Table 5.3 presents the ANOVA table for SWR, whereas Table 5.4 provides the ANOVA table for CoF. The total sum of squared deviations is employed to calculate the proportion of contribution from each input element, which is subsequently utilized to investigate the relationship between the input characteristics and abrasive wear characteristics.

Table 5.3 ANOVA for specific wear rate (mm^3/Nm)

Source	DF	Seq. SS	Adj. MS	F_{cal}	% of contribution
% Reinforcement	3	0.000005	0.000002	10.89	15.625
Applied Load (N)	3	0.000013	0.000002	28.49	40.625
Sliding Distance (m)	3	0.000013	0.000004	27.31	40.625
Error	6	0.000001	0.000000		3.125
Total	15	0.000032			100

Table 5.4 ANOVA for coefficient of friction (μ)

Source	DF	Seq. SS	Adj. MS	F_{cal}	% of contribution
% Reinforcement	3	0.003059	0.001020	37.61	59.168
Applied Load (N)	3	0.001306	0.000435	16.07	25.261
Sliding Distance (m)	3	0.000642	0.000214	7.89	12.418
Error	6	0.000163	0.000027		3.153
Total	15	0.00517			100

The percentage of ZBP filler, applied load, and sliding distance have been determined to be significant factors, as indicated by the statistically derived F-values for SWR and CoF, which exceed the F-table values. The ANOVA results indicate that the F values for the SWR in the experimental data are 10.89, 28.49, and 27.31, respectively. These values are higher than the F-test value obtained from the statistical analysis, which is 4.45. The study of SWR focused on several key aspects. The most significant component was the level of ZBP reinforcement, which contributed to 15.625% of the whole results. Following closely after were the applied load and the sliding distance, which accounted for 40.625% of the total. DOF is for degrees of freedom. Adj. SS represents adjacent sum of squares. MS stands for mean sum of squares.

The experimental F values for the COF are 37.61, 16.07, and 7.89. The primary determinants observed in the study were the percentage of reinforcement (59.16%), coupled with the applied load (25.26%) and the sliding distance (12.41%). Several studies have also documented comparable findings, leading to the conclusion that the abrading distance has a positive correlation with the wear rate. Consequently, it can be inferred that the input variables of sliding distance, applied load, and reinforcement have a significant role in determining the features of abrasion wear. The coefficient of determination (R-Sq) for SWR and COF is 96.61% and 97% correspondingly.

5.6 WORN SURFACE MORPHOLOGY OF ABRADED GFRP/ZBP COMPOSITES

SEM photographs of the worn surfaces provide valuable insights into the mechanism of abrasive wear. SEM photos of the worn surfaces are shown in Figure 5.7 (a-d). Certain mechanisms, such as the process of abrasion, lead to the removal of material from the surface of the worn samples. Abrasion surfaces are becoming increasingly complicated, with a number of process leading to weight loss being discovered with each passing year. Microploughing, microcutting, microfatigue, and microcracking are some of the processes involved in the abrasive wear process, which are described in more detail below. The arrow in Figures 5.7 (a-d) represents the abrading direction.

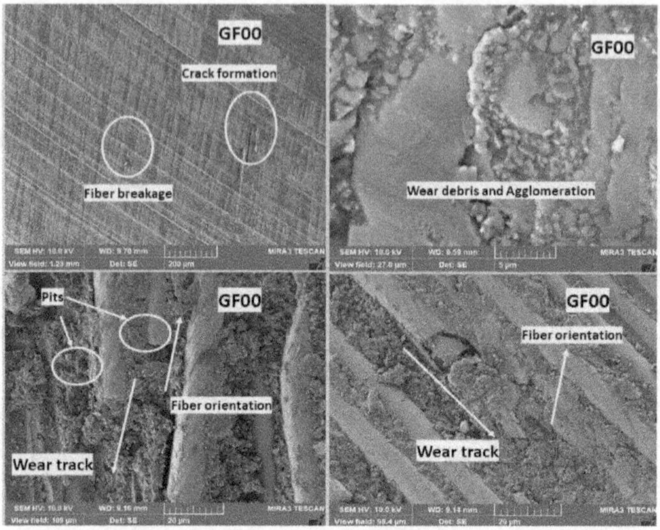

Figure 5.7 (a) SEM image of Glass fiber abraded surface

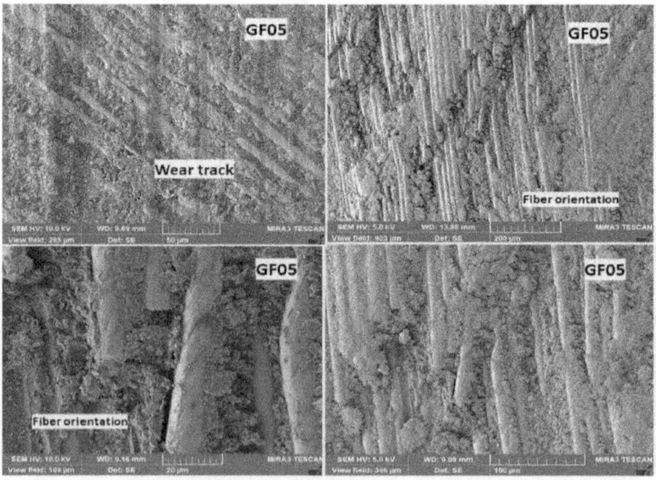

Figure 5.7 (b) SEM image of 5 wt. % reinforced ZBP in GFRP.

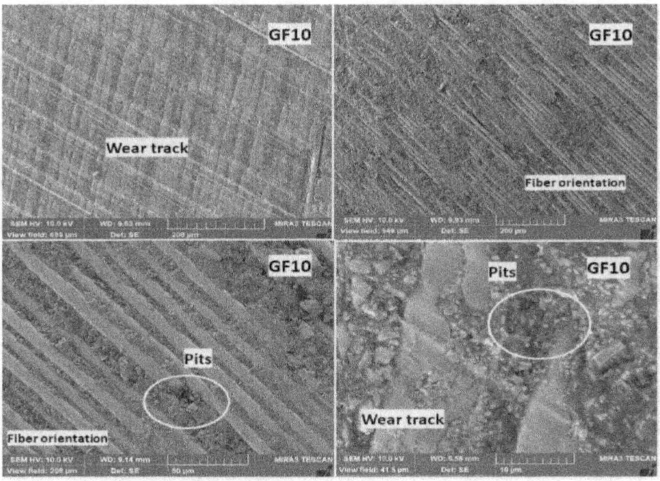

Figure 5.7 (c) SEM image of 10 wt. % reinforced ZBP in GFRP.

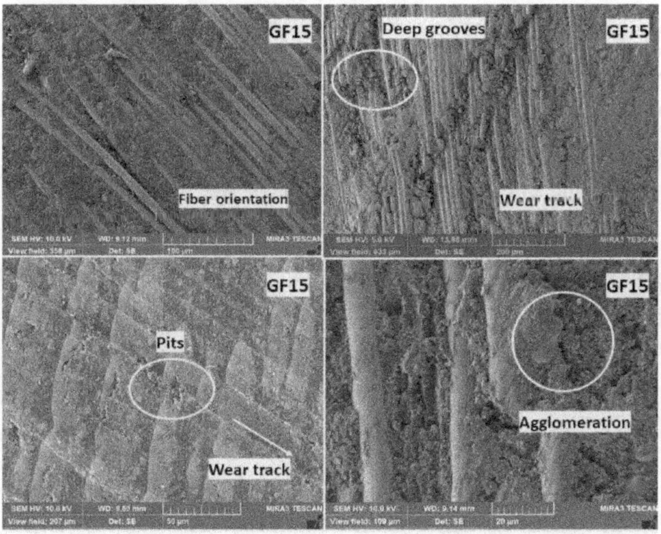

Figure 5.7 (d) SEM image of 15 wt. % reinforced ZBP in GFRP.

A SEM was used to examine the worn surface morphology of the abraded GFRP/ZBP specimens. Various mechanisms have been involved and proposed to explain the removal of material during abrasion wear. Throughout the abrasion of the specimen against the 320-grit size SiC paper at various sliding distances and loading conditions, material loss occurs. There are several mechanisms involved to identify the material removed from the surface. The wear loss in abrasion wear doesn't constitute mainly one single mechanism, and the complex of wear rate contribution is based on four major mechanisms, namely microplowing, microcracking, microcutting, and microflavor. The abrasive wear performance of GFRP is enhanced with filler reinforcement. Under abrasion, we can qualitatively identify the dominant wear mechanism from the SEM micrograph.

Figure 5.7 (a) shows a pure GFRP 10N load with a sliding distance of 100 m on a SiC 320 paper-grade abrasion wear surface. The photomicrograph clearly shows the matrix had been damaged under the low load. The wear debris formed on the surface shows the deteriorated wear resistance. Due to abrasive wear, the image also shows some tearing away of the layer from the matrix surface. This mechanism is predominantly known as microcutting and microplowing. The worn surface of the GFRP-5 wt. % ZBP composite SEM micrograph is shown in figure 5.7 (b), which was abraded against the 320 grit SiC paper under 15 N load and 100 m sliding distance. It is seen that the matrix has not been damaged much compared to the GFRP. The abrasive paper had formed shallow and wide grooves on the GFRP-5 wt. %. ZBP composite parallel to the direction of friction. The 15 N load given on the specimen over the abrasion surface showed a reduced wear mechanism.

The delamination layer formed is due to the frictional heat on the surface, where the cyclic friction occurs in the same place and causes the damage to polymers. From Figure 5.7 (c & d), as the zinc borate content increased to 10 and 15 wt.%, the grooves became narrow and shallow during high loads such as 20 N and 25 N and a sliding distance of 100 m. The mechanism at the start of abrasive wear on GFRP/ZBP composite was microcutting. The abrasive SiC paper has a grit size of 320, where the individual grains remove the material by penetrating deeply into the surface through a microcutting process.

There is some wear debris that is found on the surface, which reduces the resistance against the wear. Subsequently, the plastic deformation of the polymer matrix tends to start forming furrows, but no swarf tendrils were present on the abraded surface. There is no splitting or tearing away of layers, making it more resistant to abrasion. In 5 wt. % reinforcement, the ploughing mechanism didn't dominate as it was seen in 10 wt. %. The deep groove that had been formed in the 10 wt. % reinforcement of ZBP shows the wear mechanism of fatigue wear and abrasive wear. The ZBP, having the greatest compression strength and temperature resistance factor, has shown its effect as particulate reinforced powder in the GFRP matrix.

The reduced furrows and smooth ploughs were seen at 15 wt. % reinforcement of ZBP. As the sliding distance contributed to a major factor in wear, a thin film will be formed over the SiC paper after multiple passes during disc rotations. That layer is the smooth transfer layer, which is formed with the cluttering of wear debris between the grit particles in SiC paper, thus reducing the wear due to the reduction in effectiveness of abrasive grain penetration.

As shown in Figure 5.5 (d), ductile fracture is the primary failure mechanism, and the degree of matrix damage is significant when compared to all other made GFRP/ZBP composites. Surface fatigue caused by repeated abrasion may be seen in Figures 5.5 (c) and (d), where wide, deep grooves can be seen running the length of the worn surface in the longitudinal direction.

Figure 5.5 (b) shows micro-cracking and micro-ploughing on the surface of 5 wt. % GFRP/ZBP composites, which is also attributable to reduced heat stability. The ZBP particle tends to become free and be eliminated as wear debris where the network of fractures crosses. In other words, a lower matrix ductility causes the matrix to deform through surface disintegration, resulting in a higher degree of wear at 5 wt% GFRP/ZBP composites.

The dominant wear mechanisms that are observed in the worn surface morphology of abraded GFRP/ZBP composites are micro-cracking and micro-ploughing. The worn surface morphology of abraded GFRP/ZBP composites mainly depends upon the percentage of ZBP particulates (wt %), sliding distance (m), and applied load (N). The trend that is found in the values of CoF and SWR is found in good correlation with the worn surface morphology.

5.7 SUMMARY

This chapter presents the process of abrasive wear for GFRP/ZBP composites, as well as the SWR and COF for the composite specimens that were created utilizing the compression moulding process. The surface morphology of the composite specimens is examined in depth using a SEM picture.

CHAPTER 6

FABRICATION AND TESTING OF COMPOSITE PIPES

6.1 INTRODUCTION

GFRP pipes have become a versatile and innovative solution in different industrial applications, providing a convincing alternative to conventional metal pipes. GFRP possesses distinctive characteristics, including resistance to corrosion, low weight, and effortless installation, which have led to its growing popularity in industries such as water treatment, chemical processing, and oil & gas.

The filament winding technique is a crucial manufacturing process employed to make GFRP pipes. This sophisticated technique enables meticulous manipulation of the positioning and alignment of the strengthening glass fibers, leading to a meticulously designed and adaptable pipe configuration.

The filament winding method involves the continuous winding of glass fibers or rovings that have been impregnated with resin onto a revolving mandrel. This is done in a specific pattern to form a composite structure with many layers. This method allows for the production of GFRP pipes with customized mechanical characteristics, such as superior strength-to-weight ratios and exceptional resistance to internal pressure.

6.2 DEVELOPMENT OF FRP PIPES

Ultrasonication was used in this process of fabricating FRP pipes in order to achieve uniform dispersion in the epoxy resin. Nano zinc Borate was included in the epoxy matrix, which was composed of Araldite LY556. This was done in order to achieve the desired effect of producing uniform dispersion. Previously, sonication was a commonly used method to consistently disperse nanofillers in the matrix resin. A glass beaker was utilized to quantify the necessary amount of nano ZBP, which was then added to a mixture of epoxy glue. The percentages were 5%, 10%, and 15% by weight. To blend the various elements, a whisker was used. Subsequently, the combination underwent further agitation using sound waves emitted by a probe operating at a frequency of 20 kHz. To create a nanogel solution, the TETA hardener (HY951) was added to the mixture after the nanogel had been well mixed. Both the epoxy resin and hardener were obtained from Huntsman Corporation. The glass fibers were obtained from Urja Products Private Limited, located in Ahmedabad, India.

The ZBP, with an average size of 50 nm, were acquired from Sigma Aldrich, India. The nanoparticles were used as fillers. ZBP has a melting point of 980 °C and possesses a density of 3.64 g/cm^3. In order to blend the nanoparticles that had been incorporated into the epoxy for a period of twenty minutes, an ultrasonic mixer was utilized. During the entire procedure of integrating the nanoparticles and epoxy mixture, a thermometer was employed to monitor the temperature of the mixture and make certain that it stayed under control of the temperature. Ultimately, the hardener was integrated into the combination of epoxy and nanoparticles, and it was mechanically blended using a stirrer. Upon comprehensive examination, it was ascertained that this specific time duration was necessary for the nanoparticles to be uniformly dispersed throughout the epoxy.

6.3 STRUCTURAL GEOMETRY

The structural composition of various specimens involved in the present experimental studies are a cylindrical tubular structure with dimensions as Length = 450 mm; Outer diameter = 63 mm; Wall thickness = 2 mm; the three-dimensional sketch of the specimen is represented in Figure 6.1.

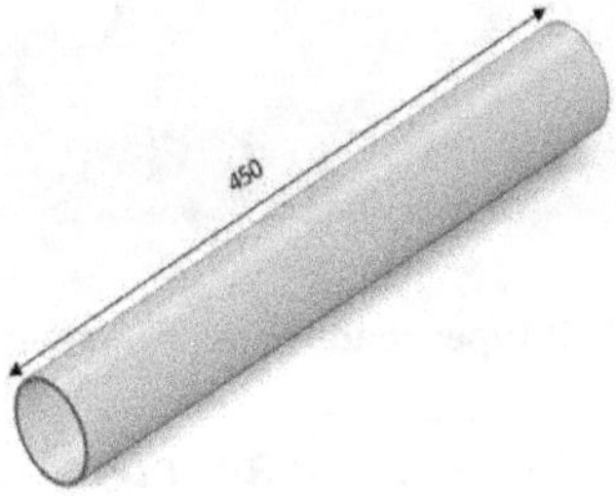

Figure 6.1 Pipe Geometry

Table 6.1 describes the nomenclature of the various materials involved in the present experimental study. Figure 6.2 shows the filament winding machine used for the fabrication of GFRP Pipes.

Table 6.1 Designation and Composition of fabricated Glass Fiber/Zinc borate composite

Designation	Composite composition by wt. %
GF	Glass Fiber
ER	Epoxy resin
ZBP	Zinc Borate Particles
GF00	50 % GF + 50% ER
GF05	50% GF + 5 wt. % ZNB + 45% ER
GF10	50% GF + 10 wt. % ZNB + 40% ER
GF15	50% GF + 15 wt. % ZNB + 35 % ER

Figure 6.2 GFRP Pipes made using Filament winding machine

6.4 PROCEDURE FOR GFRP PIPE MAKING WITH AND WITHOUT ZINC BORATE

It was determined that the filament winding process would be the most effective way to manufacture GFRP/ZBP pipes. According to this technique, the fibers are first immersed in a resin that has been combined with nano gel ZBP fillers. After that, the fibers are passed through a ZBP resin bath and then twisted around the turning mandrel in a particular pattern or at an angle of 45°.

On the other hand, in this specific method, all of the procedures were carried out with the assistance of a filament winding machine. The filaments that were spun around the mandrel were positioned at an angle of 45°, which ensured that the fabrication was carried out precisely and with perfection. The production of GFRP/ZBP pipes required the use of mandrels that were made of steel or PVC pipes of varied diameters. First, a layer of mould oil was applied to the surface of the mandrel that was exposed to the sun. After that, the axial orientation of the mandrel was properly aligned with the polyethylene strips that were being employed. At long last, the mandrel was wrapped in a polyethylene sheet that was securely taped around it. The process of demoulding the GFRP/ZBP pipe was significantly eased by the incorporation of mould oil and polyethylene strips, which was then followed by the curing of the GFRP composite to complete the process.

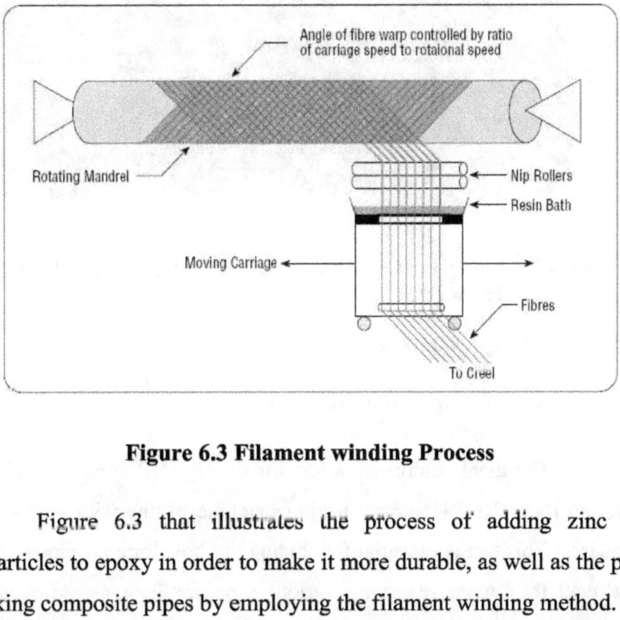

Figure 6.3 Filament winding Process

Figure 6.3 that illustrates the process of adding zinc borate nanoparticles to epoxy in order to make it more durable, as well as the process of making composite pipes by employing the filament winding method.

The composite pipe samples seen in Figure 6.4 were manufactured by employing the filament winding technique. Each layer of the pipe was composed of six layers, and the winding angle was set at ± 45°. All of the pipes have an inside diameter of 59 mm and a wall thickness of 2 mm. When the pipe that had been impregnated reached the proper width, it was taken from the mandrel. Next, it was cured at a temperature of 100°C for a period of 150 minutes.

Figure 6.4 Fabricated samples of GFRP composites with 0, 5, 10, and 15 % ZNB

6.5 COMPRESSION TESTING

6.5.1 Axial Compression Behavior

The axial compression test for GFRP/ZNB composite pipe is done using an INSTRON Universal testing machine in line with ASTM D695–15 standards. This test is essential for finding out how long oil pipe lines last and how well they work regarding various pipe applications. Applications that operate in the axial direction, such as propulsion and the movement of gasses, among others, perform very well when they are produced using the filament winding method. Because of the curvature of the fiber, the load that is applied in the radial direction has a tendency to undergo buckling or deformation. The load is applied in a direction that is perpendicular to the yarn's winding direction. One of the parameters that is utilized in the process of evaluating the effectiveness of the Zinc Borate impact in the GFRP pipe is known as the Specific energy absorption (SEA).

It provides a representation of the crashworthiness behavior as well as the energy absorption capability of the GFRP/ZBP composite pipe that was manufactured. Adding 15 weight percent of zinc borate particles resulted in the maximum energy absorption, which was 5.33 J/g. This was followed by adding 10 weight percent of zinc borate particles. The experimental results of axial compression testing of unfilled GFR pipes and filled GFR pipes are tabulated in Table 6.2.

Table 6.2 Compression tested values of GFRP/ZBP specimens

Parameter	GF00	GF05	GF10	GF15
Compressive strength (MPa)	89.322	95.034	106.527	102.457
Compressive modulus (GPa)	2.86	2.93	3.36	3.16
Pmax (kN)	34.12	37.11	36.61	34.56
Pmean (kN)	24	27	28	26
Mass (g)	45.11	42	41.77	41.6
EA (J)	335	343	401	398
SEA (J/g)	4.96	4.98	5.31	5.33
CFE	0.70	0.73	0.76	0.75

As a result of the greater inter layer bonding strength with the resin matrix, the GF10 specimen was found to have the highest compressive strength, which was measured at 106 MPa as shown in Figure 6.6. This represents a 16% increase in comparison to the GF00 specimen. 89.32 MPa was the compressive strength that was generated by the composite made of glass fiber filament windings that did not contain any filler. the GF15, which demonstrated a 12.7% improvement in comparison to the GF00 specimen and a 3.3% decrease in comparison to the GF10 specimen, which had the highest compressive strength. In Figure 6.5 and Figure 6.7, respectively, the impact of crushing force is shown in phases under axial and radial directions, respectively. All of the GFRP/ZBP specimens were subjected to the same pattern of crushing action. GF10 has a compression modulus that is greater than that of any of the other specimens that were evaluated.

The ratio of the highest load to the mean load that was measured from the Load vs. Displacement curve (Figure 6.8) is the equation that represents the crush force efficiency. With a crush force efficiency of 76%, GF10 was the material that achieved the highest level, followed by GF15, GF05, and GF00. The nano ZBP-influenced GFRP had demonstrated an increase in the efficiency of the crash force. The nano zinc borate particles that were put in with the epoxy resin had resulted in improved bonding strength as well as resistance to compression. Additionally, the addition of fifteen weight percent of ZBP resulted in a rise in brittleness, as well as the formation of certain fiber cracks, which led to the breakage of the fibers during the compression failure.

The stress that is passed from the fiber to the matrix had an effect on this compression strength. This was demonstrated by the crushing load, which showed a sudden increase in its peak and then decreased down in a limited range, resulting in unstable buckling deformation in the GFRP/ZBP composite specimen. The images that are displayed in Figures 6.5 provide evidence of the damaged GFRP/ZBP composite specimens.

These photographs demonstrate that the buckling routes had occurred between the transverse cracks and the undamaged zones. Because of the increased compression caused by the cross head, the fibers begin to fracture, transverse cracks appear, and the early fractures surrounding the wall to induce shearing of the fibers result in plastic deformation, which demonstrates a better energy absorption ability at GF10. It is generally agreed upon that these findings are consistent with the observations made by a number of different studies

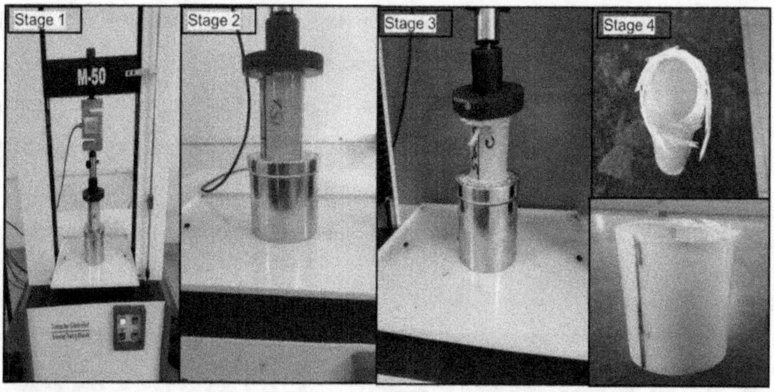

Figure 6.5 Stages of GFRP/ZBP composite under axial compression loading

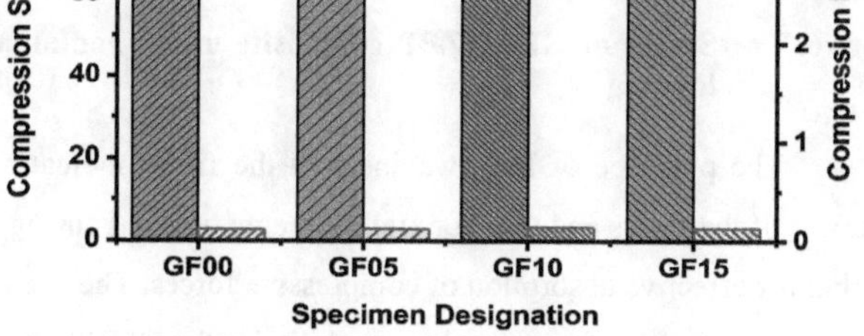

Figure 6.6 Compression strength and compression modulus of GFRP/ZBP pipe

6.5.2 Radial Compression Behavior

The specimen containing 10 wt. % of ZBP in GFRP exhibited the highest compression behavior compared to the other examined specimens, as shown by the results of the radial compression test. Compression failure often arises when materials are unable to handle the load they are subjected to. The failures are depicted sequentially in Figure 6.7, clearly illustrating the progression of compression stress resulting in buckling and collapse of the GFRP/ZBP specimens. The interlaminar failure is caused by the load exerted on the layers of filament winding, resulting in damage to the matrix and fracture of the fibers.

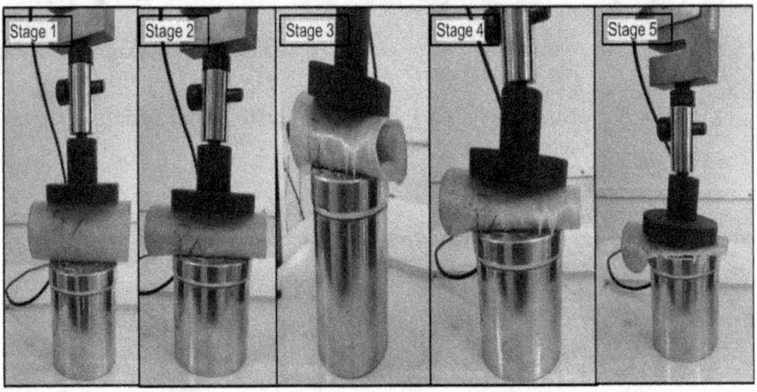

Figure 6.7 Stages of GFRP/ZBP composite under radial compression loading

The presence of little waviness in the fibers indicates the correct alignment of the fibers and the thorough impregnation and curing of the resin, allowing for effective absorption of compressive forces. The use of nano ZBP filler enhanced the energy absorption capability of the GFRP pipes. Figure 6.8 demonstrates that the specimens undergo elastic elongation until reaching the maximum crushing force, at which point plastic deformation begins. The addition of 15 wt. % of ZBP did not result in significant improvement, as evidenced by the axial compression test.

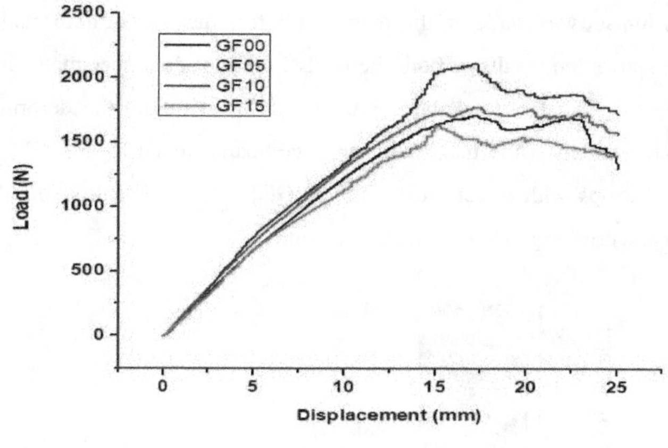

Figure 6.8 Radial compression test results of fabricated specimens

The presence of brittle fiber fracture and failure of the matrix under lower compression load suggests that the delamination of fibers and inadequate interlocking between the matrix and reinforcement may be responsible. The magnitudes of GF10 and GF15 findings are directly proportional to SEA capacity, which is enhanced by the increased presence of nano fillers in GFRP. The crash force efficiency is greater in GF15, with a value of 86%, compared to GF10 (75%), GF05 (78%), and GF00 (72%). The results obtained were comparable to the findings investigated by Praveen Kumar *et al.* (2022).

6.6 DROP WEIGHT IMPACT TESTING

The concentrated impact of GFRP/ZBP pipe specimens were investigated in two methods of placing the specimen for drop weight impact a) Axial Impact b) Transverse impact.

In axial loading, the drop weight impact was performed in two loading conditions of 50 N and 60 N from a height of 2 m and with a velocity of drop of 6.32 m/s. The test was evaluated based on the number of times the axial impact was made on the specimen before the start of deformation. The table shows the results of both the loads for all the four specimens. From the Table 6.3, at 50 N load the GF10 took 5 drops before the deformation is initiated which is the maximum when compared to GF15 and GF05 which took 4 drops with direct crushing. The GF00 shows delamination at the 3rd drop and crushing behavior at the 4th drop.

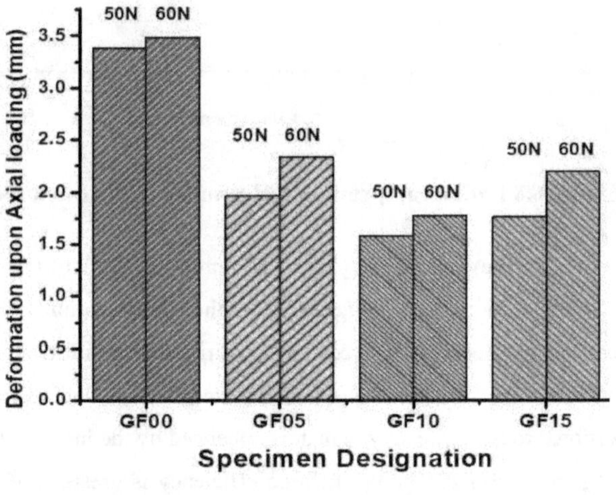

Figure 6.9 Low-velocity impact test of GFRP/ZBP composite pipe in axial direction

From the Figure 6.9, the influence of nano ZBP on the impact behavior with the amount of deformation upon axial loading is studied and it's due to the fiber bonding with the matrix and the specific energy absorption. The interlocking of glass fiber with the nano ZBP shows maximum resistance to deformation at 10 wt. % and again the deformation increases at 15 wt % addition for both the loads.

Table 6.3 Impact test values of GFRP/ZBP specimens

Sp. No.	I_L (mm)	I.D (mm)	T (mm)	P (N)	H (m)	t (Sec)	V (m/s)	n	F_L (mm)	Type of deformation
GF00	250	59	2	50	2	0.41	6.32	3	246.62	Delamination & Crushing
GF05	250	59	2	50	2	0.42	6.32	4	248.04	Crushing
GF10	250	59	2	50	2	0.40	6.32	5	248.42	Crushing
GF15	250	59	2	50	2	0.43	6.32	4	248.24	Crushing
GF00	250	59	2	60	2	0.40	6.32	2	246.52	Delamination & Crushing
GF05	250	59	2	60	2	0.42	6.32	3	247.47	Crushing
GF10	250	59	2	60	2	0.41	6.32	4	248.23	Crushing
GF15	250	59	2	60	2	0.43	6.32	3	247.51	Crushing

6.6.1 Deformation of Specimen Upon Drop Weight Impact in Radial Direction

The low velocity impact test is performed on the filament wound pipes in transverse direction to find the impact effect of GFRP/ZBP specimens. This test will help in understanding the bursting pressure of GFRP pipes as said by Gemi *et al.* (2018). The test was performed at 3 m/s velocity and from a height of 2 m. the impact is subjected to a point on the top layer of the GFRP/ZBP specimens, From the Figure 6.10 we can find the delamination occurred with the amount of drop impact. The layers get delaminated, interlayer matrix damage and the damage of fiber breakage is seen on the impact area. From the area of delamination as shown in stages, the least damage is obtained by GF10 specimen and matrix cracks is minimum showing the resistance offered by the 10 wt. % addition of ZBP in GFRP. The GF00 shows a larger area of delamination and resistance to impact is less. Similar type of study results was seen in the investigation done by DilekSoyluGemi *et al.* (2022).

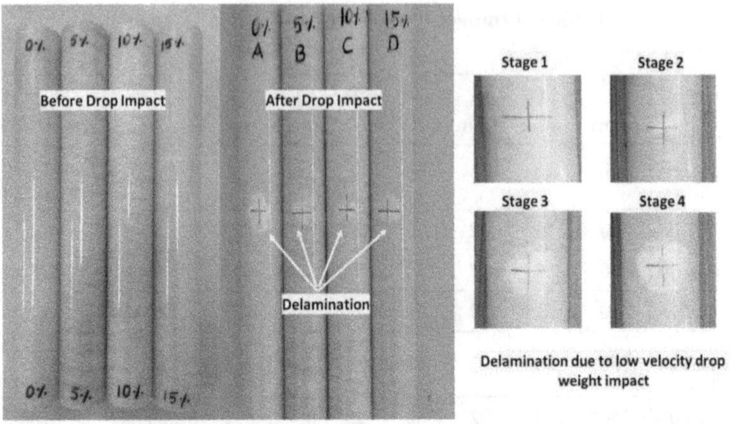

Figure 6.10 Stages of GFRP/ZBP composite under low velocity drop weight impact test

6.7 HYDROSTATIC PRESSURE TESTING

The hydrostatic pressure testing is experimented using hydraulic fluid and pneumatic air to evaluate the pressure test for pipes as shown in Figures 6.11 and 6.12. ASTM D1599 procedure is adapted to find the failure pressure of the GFRP/ZBP pipes. The pipe specimens were subjected to an internal pressure and the hydraulic water pressure is raised uniformly till the failure happens in the pipe. A hydrostatic pressure pump is utilized and the pressure is kept on increasing into the specimen with a non-return valve to prevent the return flow and pressure drop while performing the test. The Figure 6.13 shows the hydraulic burst pressure for various ZBP addition in GFRP pipe.

The GF10 shows the maximum resistance to failure among all specimens with 30% enhancement due to ZBP addition. The GF15 shown a reduction of 3% compared to GF10, due to the decrease in the fiber matrix interlock bonding. The crack propagation formed by the bursting pressure was whitening or delamination of fiber and matrix, crack initiation, crack propagation and burst leakage. The GFR pipes underwent simultaneous deformation in both the radial and axial axes when subjected to internal pressure.

The deformations were quantified utilizing a pair of linear position sensors manufactured by Novotechnik. During the internal pressure testing, the radial displacement of the GFR pipes was measured from the outside surface of the GFR pipes at the midpoint of its axial length. The data logger received the displacements that took place in both the radial and axial directions of the pipes.

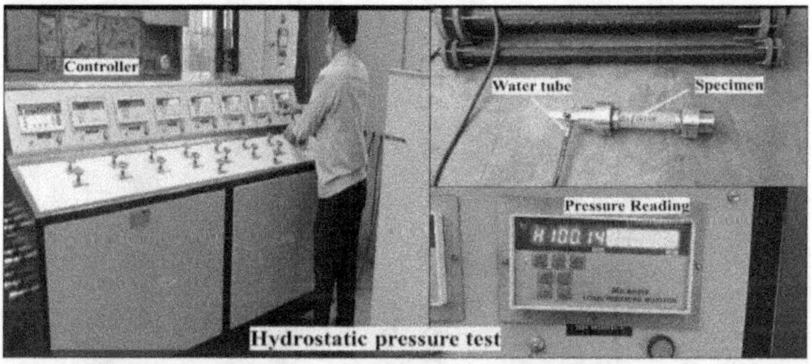

Figure 6.11 Setup of Hydrostatic burst pressure test

The GFRP pipes are placed with a valve at one end and an end cap at another end as shown in Figure 6.12 The pressurized pneumatic air or gas is passed into one side of the GFRP pipe and made to fill inside. The pressure is kept constant and for a time period of 10 min. This creates a high-pressure environment inside the pipe and it will evaluate the pipes reliability, functionality, safety and leak tightness of the pressure system before being put into practical applications. This test is conducted in accordance with industry safety standards.

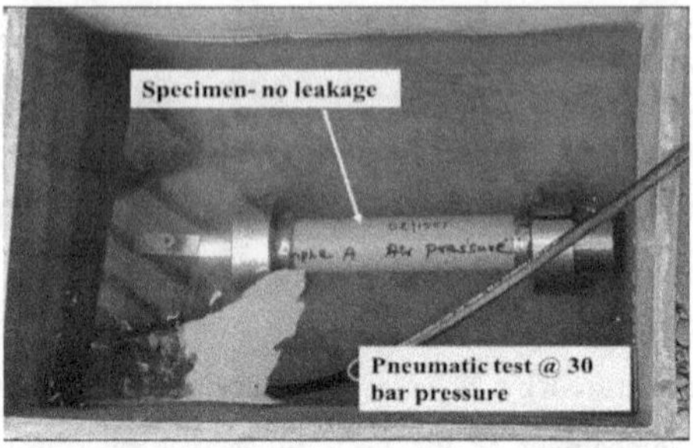

Figure 6.12 Setup of Pneumatic pressure test

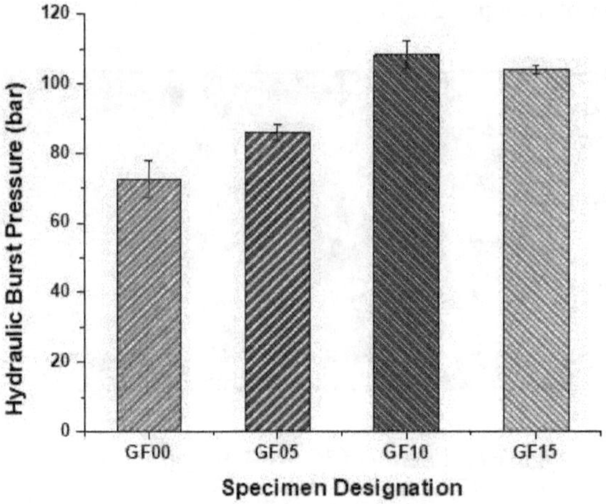

Figure 6.13 Hydraulic burst pressure test of GFRP/ZBP composite.

The hydrostatic pneumatic test is examined by with limitations industrial safety standards, where the specimens of GFRP/ZBP were clamped up and filled the pipe specimens with 30 bar pressure and specimens were dropped inside water for 10 min. This test showed no sign of failure in all the specimens examined. This provides the GFRP/ZBP pipes safety, reliability, functionality and leak tightness of the pressure system.

6.8 FRACTOGRAPHY RESULTS

To analyze the failure modes and levels of damage occurred at different specimens, SEM analysis was taken on some cut specimens of the tested GFRP/ZBP pipes. The SEM images in Figure 6.14 (a – f) shows the different specimens and fracture during axial and radial compression. The dominant failure modes were the delamination, buckling/ fiber fracture, voids in interfacial cracks propagation. The Figure 6.14 (a – f) shows the voids and fiber fracture that is causing the delamination and whitening of impacted area and initiate the failure mode.

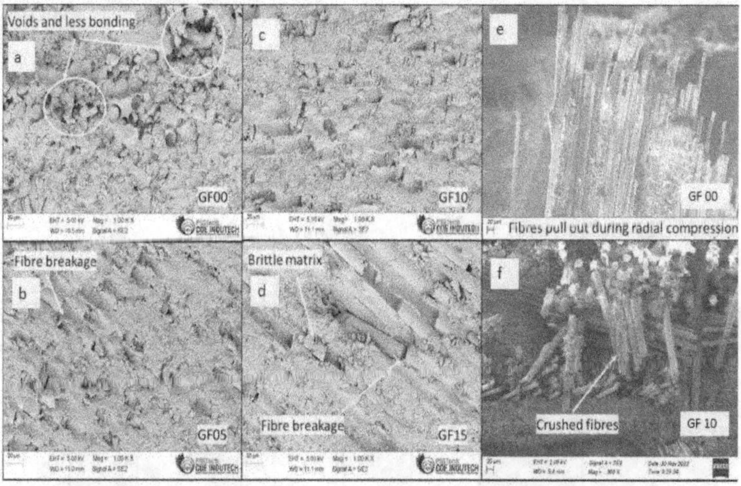

Figure 6.14 (a – d). Fractography of GFRP/ZBP composite pipes under axial compression and (e – f). Fractography of GFRP/ZBP composite pipes under radial compression.

The GF10 which had given the best results in all the studies shows the dense packing of GFRP with nano gel ZBP resin coating where the fiber matrix interlocking is more to transfer of applied stress from matrix to fiber. GF00 which is less dense comparatively showed reduced resistance in the performed tests as the fiber matrix bonding is not good. The interlaminar failure, interlaminar damage, matrix damage and buckling were observed in the GFRP pipes manufacture through filament winding process. These failure mechanisms will show the significance of the GFRP/ZBP composite pipe for various structural applications. Similar type of SEM results were observed by the study performed by Mustaffa *et al.* (2023).

6.9 VIBRATION TEST

The Figure 6.15 depicts the damping factors of various specimens of a GFRP composite reinforced with Zinc Borate. The composite is fixed at both ends and tested under three different modes of vibration: Mode 1, Mode 2, and Mode 3.

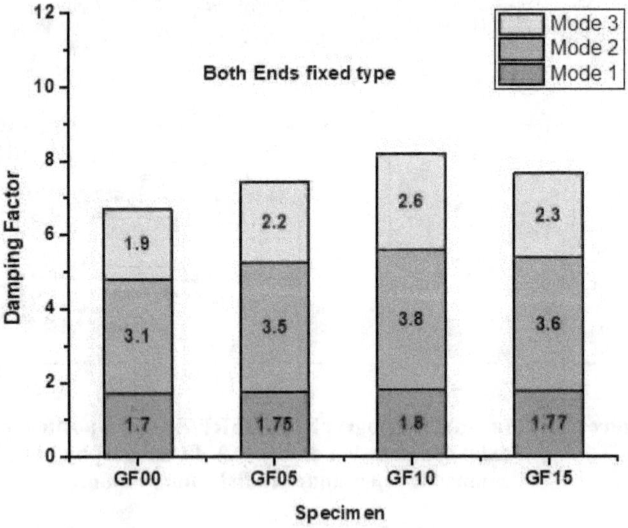

Figure 6.15 Damping factors for GFRP pipe when both ends are fixed

The damping factors show a consistent rise from GF00 to GF10, followed by a minor decline in GF15. This indicates that the addition of Zinc Borate may improve the damping characteristics up to a specific threshold (GF10), after which there is a minor decrease. Mode 2 often exhibits the largest damping factor among all specimens. Mode 1 exhibits the lowest damping factor among all specimens.

Mode 3 exhibits a moderate level of dampening. Adding Zinc Borate to GFRP composites enhances their damping properties, specifically in Mode 2 vibration. The enhancement reaches its maximum for the GF10 sample, followed by a minor decrease in the damping factor for the GF15 sample. This data can be utilized to enhance the quantity of Zinc Borate reinforcement in order to attain optimal damping performance in real-world scenarios.

The Figure 6.16 depicts the damping factors of various specimens of a cantilever-type GFRP composite reinforced with Zinc Borate in three different modes of vibration (Mode 1, Mode 2, and Mode 3).

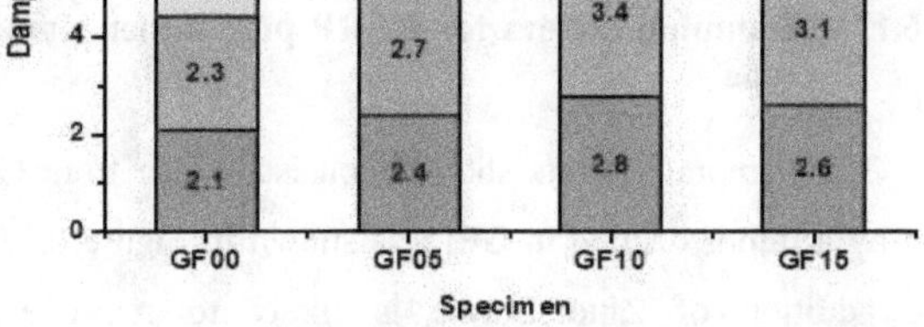

Figure 6.16 Damping factors for GFRP pipe under both cantilever-type

Typically, Mode 2 has the largest damping factor among all specimens. Mode 1 has the least amount of damping compared to all other specimens. Mode 3 exhibits a moderate level of dampening.

Adding Zinc Borate to GFRP composites enhances their damping properties, especially in Mode 2 vibration. The enhancement reaches its maximum for the GF10 sample, followed by a minor decrease in the damping factor for the GF15 sample. This data can be utilized to improve the quantity of Zinc Borate reinforcement in order to attain the most effective damping performance in real-world scenarios.

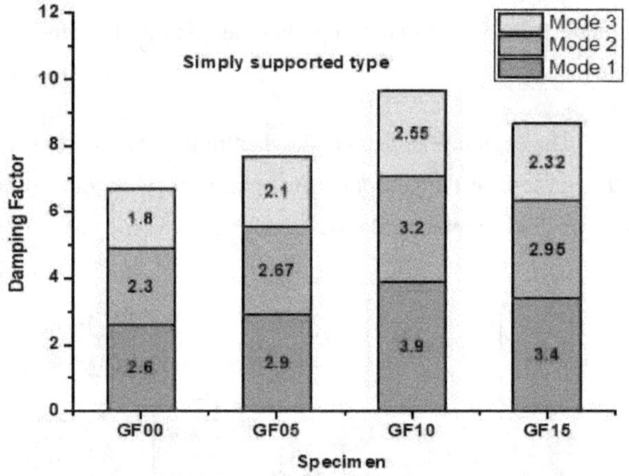

Figure 6.17 Damping factors for GFRP pipe under simply supported-type

The damping factors show a consistent rise from GF00 to GF10, followed by a minor decline in GF15 as shown in Figure 6.17. This implies that the addition of Zinc Borate is likely to improve the damping characteristics until a specific threshold (GF10), after which there may be a little decline.

Adding Zinc Borate to GFRP composites enhances their damping properties, especially in the primary vibration mode. The enhancement reaches its maximum for the GF10 sample, followed by a minor decrease in the damping factor for the GF15 sample. This data can be utilized to enhance the quantity of Zinc Borate reinforcement in order to attain optimal damping performance in real-world scenarios.

6.10 WATER ABSORPTION TEST FOR GFRP COMPOSITE PIPE

Water absorption test was carried out to regulate the amount of water to be absorbed by the specimen. For this experiments ASTM standard D570 was selected and then specimens were fabricated as per the standard dimension. Initially the specimens were dried with oven to remove the moisture content.

Further the specimens are got weighed and they are dipped into various types of water like distilled water, normal water and salt water. The material is then emerged in water at agreed upon conditions, often 23°C. These experiments were taken at a time interval of 48 hours. After 48 hours the specimen were taken out, cleaned properly and weighed. The weight of the specimen was getting increased. The procedure can be repeated for the 48 hours for all 3 the specimens under 3 different water environment.

From the table 6.4 and 6.5 it's declared that, all the composite materials specimens were absorbed little or negligible amount of water under all three different water solutions.

The newly made encompassing glass fiber, epoxy resin and zinc borate has been intended in detail. Experimental results were correlated with existing literature results (Where specimen made up only by GFRP) and found better material properties for these newly made composite materials (GFRP with ZBP) for water absorption environments.

Table 6.4 Before water absorption test

Specimen number	Water Solution type	Weight in grams (m_i)
GF1	Normal water	312
GF2	Normal water	364
GF3	Normal water	416
GF4	Normal water	468
GF1	Salt water	312
GF2	Salt water	364
GF3	Salt water	416
GF4	Salt water	468
GF1	Distilled water	312
GF2	Distilled water	364
GF3	Distilled water	416
GF4	Distilled water	468

The Mathematical model for water absorption test is shown in equation 6.1,

$$M = (m_o - m_i)/m_i \quad (6.1)$$

Where m_o = final mass (g)

m_i = initial mass (g)

M = percentage of water absorption

Table 6.5 After water absorption test (Time interval 48 hours)

Specimen number	Water Solution type	Weight in grams (m_o)	Difference (grams) (m_0 - m_i)	Water absorption M (%)
GF1	Normal water	315	3	0.95
GF2	Normal water	368	4	1.09
GF3	Normal water	418	2	0.48
GF4	Normal water	472	4	0.85
GF1	Salt water	314	2	0.64
GF2	Salt water	367	3	0.82
GF3	Salt water	417	1	0.24
GF4	Salt water	470	2	0.43
GF1	Distilled water	316	4	1.27
GF2	Distilled water	369	5	1.36
GF3	Distilled water	418	2	0.48
GF4	Distilled water	471	3	0.64

6.11 SUMMARY

The fabrication of GFRP/ZBP composite pipes through filament winding process, and experimental results of GFRP/ZBP composite pipes under compression, drop weight impact and hydrostatic pressure loading was presented in this chapter. The fractography of the composite pipes with a SEM image is also explored in detail.

www.ingramcontent.com/pod-product-compliance
Lightning Source LLC
LaVergne TN
LVHW010724061126
851857LV00014B/1336